Occupational Health Matters in General Practice

Ruth Chambers

Stephen Moore

Gordon Parker

and

Andy Slovak

Staffordshire
UNIVERSITY

Radcliffe Medical Press

©2001 Ruth Chambers, Stephen Moore, Gordon Parker and Andy Slovak
Illustrations ©2001 Dave Colton www.cartoonist.net

Radcliffe Medical Press Ltd
18 Marcham Road, Abingdon, Oxon OX14 1AA

British Library Cataloguing in Publication Data

A catalogue record for this book is available from the British Library.

ISBN 1 85775 463 8

Typeset by Advance Typesetting Ltd, Oxfordshire
Printed and bound by TJ International Ltd, Padstow, Cornwall

Contents

Preface

'Good occupational health practice ... should lead, in the longer term, to positive outcomes for workers and businesses alike, in terms of a good quality of life inside and outside work, the social and material advantages of work, reduced sickness absence, higher productivity, a good, responsible image for individual businesses and greater national wealth creation.'[1]

Every year 2.2 million people suffer from ill-health caused or made worse by their work. Thirteen million working days are lost as people take time off because work has made them ill. General practitioners and others working in primary care see the effects of work-related illnesses and injuries on their patients every day.

Many general practitioners have limited training in occupational health. Many are unfamiliar with the requirements of relevant health at work legislation; few apply Health and Safety policies in their workplaces. Patients consulting general practitioners because they are suffering from work-related ill-health do not always receive the best advice if GPs and practice nurses are inexperienced in occupational medicine.

This book has been written for GPs in their roles as employers and medical advisers to the public, but will be useful to others in the primary care team too. Each chapter ends with interactive exercises to help readers think through the issues and apply what they have learnt to their own practice. The interactive material is suitable for practice nurses and practice managers, as well as general practitioners and can be used for either continuing professional development (CPD) for individual GPs, or as a template for a personal development plan, or a practice personal and professional development plan. Worked examples of personal development plans and practice personal and professional development plans offer models for continuing professional development.

All sections of this book are suitable for individual GPs and practice nurses. Practice managers should use the material to improve the occupational health-care of the GPs and their staff; and health and safety in the practice. This multi-professional approach fits with the changes to continuing professional development in primary care where learning should become a practice-based activity that is relevant to service needs.

An option: using the material in the book for a distance learning programme

You might choose to undertake some or all of the 'Reflection exercises' at the end of some or every chapter by yourself or you might work through the material with a colleague in your practice or compare notes with someone from another practice. The answers to the reflection exercises can be deduced from the preceding text or in further reading.

The time you might expect to take to complete the reflection exercises over and above the reading required is given for each individual exercise; the exercises for the first seven chapters should take about 30 hours in all. In addition, the time estimates for undertaking a personal development plan or practice personal and professional development plan are given in Chapters Eight and Nine – but the actual time you take will depend on what prior work you have done, the extent to which you tie in the reflection exercises given here as your preliminary work to gather baseline information about current performance and shape the plans, and how ambitious your plans are.

You may prefer to use the book as a resource for self-directed learning; and if you do progress to drawing up a personal or practice-based learning plan focusing on occupational health, apply to have your plan(s) accredited by your local GP tutor, or through whatever system is in place in your locality. But you might choose to invite external feedback on your completed reflection exercises, or draft learning plans. If so, you can register for a distance learning programme at Staffordshire University where you will receive individualised feedback on your answers to the reflection exercises or completed draft learning plans with suggestions for improvements. The programme has been piloted on general practitioners who rated the material highly. This programme has been recommended for up to 40 hours of postgraduate education accreditation (PGEA) for GPs by the UK National Accreditation Panel (25 hours disease management and 15 hours service management); the recommendation has been passed on to all Directors of Postgraduate General Practice Education. An individual GP should apply to their Director of Postgraduate General Practice Education for 'PGEA'.

The aims of the distance learning programme are to:

- introduce the basics of occupational medicine and commonly encountered occupational health problems
- introduce some of the terminology
- help you to understand the roles and responsibilities of members of the primary healthcare team in relation to occupational health
- help you to understand the role of occupational health services

- help you to integrate that learning into your everyday general practice work so that you recognise the effects of health on work; and the effects of work on health – in relation to your patients, and colleagues in the NHS
- enable you to help people to return to work safely and effectively.

Please send for details of the distance learning programme to:
Occupational Health in General Practice distance learning programme
Centre for Health Policy and Practice, School of Health
Staffordshire University
Blackheath Lane
Stafford ST18 0AD
Fax: 01785 353673.

Ruth Chambers
Stephen Moore
Gordon Parker
Andy Slovak
September 2000

---

1   Occupational Health Advisory Committee (2000) *Improving Access to Occupational Health Support.* HSE Books, Sudbury.

About the authors

Ruth Chambers has been a GP for 20 years. She has undertaken a wide range of research focusing on stress and the health of doctors, health at work and the quality of healthcare. She is currently the Professor of Primary Care Development at Staffordshire University. Ruth has designed and organised many types of educational initiatives, including a long-distance learning programme, *Survival Skills for GPs*. Recently she has developed a keen interest in working with GPs, nurses and others in primary care around clinical governance and practice personal and professional development plans.

Stephen Moore left his GP partnership after 11 years to further his growing interest in occupational medicine. He has served as a member of the advisory panel to the Food and Drink Federation Occupational Health and Safety Group as well as being an expert witness in the field of work-related illness and injury. He lectured on health and safety legislation at the Universities of Manchester and Birmingham, and GP postgraduate education meetings. Stephen is currently an independent medical adviser and occupational physician to a number of companies throughout the East Midlands.

Gordon Parker was a GP principal for 5 years before training in occupational medicine with Nuclear Electric plc, PowerGen plc and Lancaster Acute and Priority NHS Trusts. As an accredited specialist, he subsequently became Head of Health and Safety Services and lecturer in occupational medicine at the University of Manchester. He is now Senior Medical Adviser to Ranks Hovis McDougall Ltd, honorary lecturer at the University of Manchester, and Training Dean for the Faculty of Occupational Medicine of the Royal College of Physicians.

Andy Slovak specialised in occupational medicine early in his medical career, starting as works medical officer at ICI Organics Division, becoming an accredited specialist in occupational medicine and progressing to his current post of Chief Medical Officer at BNFL plc in 1990. He combines this post with being Senior Lecturer in occupational medicine at the University of Manchester and consultant in occupational medicine for Reuters. Andy has a keen interest in upper limb disorders; he has a well-established academic track record, lecturing and publishing widely on occupational matters. His doctorate was on laboratory animal allergy.

Acknowledgements

This programme was developed by the authors with funding and support from the Health and Safety Executive under the guidance of a joint committee from the Faculty of Occupational Medicine of the Royal College of Physicians, the Society of Occupational Medicine, the Royal College of General Practitioners and the British Medical Association, chaired by Dr Lotte Newman CBE. The original idea for the programme was conceived by Dr Vivienne Press from the Department of Health.

We should particularly like to thank June Manson and Tony Leach from the Health and Safety Executive, GPs Susan Jenkins and Sheila McCorkindale, and GP educationalist Shake Seigel for their input into the original material.

The programme has been published with their further support and that of Dr Philip Sawney (Medical Policy Manager, Department of Social Security), Dr Jim Ford (Primary Care Division, Department of Health/NHSE), Dr Olivia Carlton (Occupational Health Policy, Department of Health) and Dr Kit Harling (ex-Dean, Faculty of Occupational Medicine of the Royal College of Physicians). We are very grateful to June Manson and Philip Sawney for checking through and advising on the manuscript.

Glossary

ACOP	Approved Code of Practice
COSHH	Control of Substances Hazardous to Health Regulations
CPD	continuing professional development
DDA	Disability Discrimination Act
DEA	Disability Employment Adviser
DSS	Department of Social Security
DVLA	Driver and Vehicle Licensing Authority
EAP	employee assistance programme
EMAS	Employment Medical Advisory Service
FEV_1	forced expiratory volume in 1 second
GHQ	General Health Questionnaire
HAVS	hand–arm vibration syndrome
HSAWA	Health and Safety at Work Act
HSC	Health and Safety Commission
HSE	Health and Safety Executive
IHR	ill-health retirement
LGV	large goods vehicle
MEL	maximum exposure limit
NIHL	noise-induced hearing loss
ODIN	Occupational Diseases Information Network
OES	occupational exposure standard
OH	occupational health
OPRA	Occupational Physicians' Reporting Activity
OSI	occupational stress indicator
PSV	public service vehicle
PEFR	peak expiratory flow rate
PPE	personal protective equipment
RADS	reactive airways dysfunction syndrome
RIDDOR	Reporting of Injuries, Diseases and Dangerous Occurrences Regulations
RSI	repetitive strain injury
RTA	road traffic accident
SMEs	small and medium enterprises
SWI	Self-reported Work Injury
SWORD	Surveillance of Work-related and Occupational Respiratory Disease
VDU	visual display unit
VWF	vibration white finger
WRULD	work-related upper limb disorder

Definitions

Hazard: is something with potential to cause harm; risk is the likelihood of that potential being realised.[1]
Risk: see 'hazard'.
Worker: all employees employed at the workplace and other employees and self-employed persons who regularly work there, e.g. agency workers and sub-contractors.[2]

References

1   Health and Safety Executive (1998) *Five Steps to Successful Health and Safety Management*. HSE Books, Sudbury.

2   Health and Safety Executive (2000) *Employee Consultation and Involvement in Health and Safety*. HSE Books, Sudbury.

CHAPTER ONE

Introduction to occupational health

Almost half the adult population of the UK is employed, and the work that people do may affect their health. During their work, some people are injured, become ill, or find it difficult to cope with work because they are ill.

In Britain in 1995–96, 1.3 million people reported ill-health that they attributed to their work.[1] A further 1 million injuries at work were reported. A total of 24.3 million working days were lost because of injuries and illness, and over 27 000 people per year had to give up their work because of ill-health. The total cost to patients and their families in lost income and extra expenditure is in the order of £5.6 billion per year. However, counting people no longer in work, a 1995 survey[2] found that the estimated prevalence of work-related illness in Great Britain in 1995 was 2 million. On this basis, the actual costs to society may be nearer £11 billion per year (based on figures[1]). Those working in general practice regularly encounter work-related problems, and may be able to intervene to reduce these health effects and financial losses.

There are two sides to occupational health:

- the effects of work and the workplace on health
- the effects of ill-health on an individual's fitness for work.

There are clinical and environmental components to most work-related health problems, together with wider issues of preventive medicine. 'Occupational health' aims to be proactive in promoting health and protecting the worker against occupationally acquired illness, rather than simply reacting to medical or environmental problems as they arise.

According to the Occupational Health Advisory Committee, the definition of occupational healthcare embraces:[3]

(a) the effect of work on health, whether through sudden injury or through long-term exposure to agents with latent effects on health, and the prevention of occupational disease through techniques which include health surveillance, ergonomics, and effective human management systems

continued overleaf

(b)   the effect of health on work, bearing in mind that good occupational health practice should address the fitness of the task for the worker, not the fitness of the worker for the task alone
(c)   rehabilitation and recovery programmes
(d)   helping the disabled to secure and retain work
(e)   managing work-related aspects of illness with potentially multifactorial causes (e.g. musculoskeletal disorders, coronary heart disease) and helping workers to make informed choices regarding lifestyle issues.

Health professionals and occupational issues

Members of the primary healthcare team have a number of different roles in relation to occupational health matters. For example, as well as general practitioners (GPs) having a professional and clinical role with their patients, they have a 'managerial' role in relation to the practice staff, visitors to the practice premises – and to themselves.

A basic knowledge of occupational health matters will underpin all the roles of the National Health Service (NHS) GP:

• professional medical adviser to patients of working age: diagnosis, management, advice on fitness for work
• provider of factual information on certain working-age patients to the Department of Social Security
• employer required to observe Health and Safety and other workplace legislation
• professional liaison point with employer or employer's medical advisers – not compulsory but part of a best practice approach.

Elements of a GP's occupational health role in relation to patients, staff and others who may use the practice and its premises[4]

1   In relation to patients:
    • early diagnosis of occupationally acquired diseases
    • management of illness caused by work or made worse by work
    • liaison with occupational health services or employers to prevent a patient's condition being exacerbated by their work, or to prevent other employees' health being affected
    • early diagnosis and assessment of conditions which may affect fitness for work

continued opposite

- advising whether or not to refrain from regular occupation
- advising on returning to work after illness or injury
- advising on ill-health retirement (general advice to patients, advice to employer or pension fund manager – as appropriate)
- advising on range of compensation and benefits.

2 In relation to staff and visitors (including patients, other healthcare workers and representatives attending the surgery premises):
- producing a health and safety policy for the practice
- identifying and assessing the risks to staff from the work of the practice
- ensuring that the identified risks are controlled, and assessing the adequacy of those controls
- advising and training staff on safe working
- obtaining appropriate advice on health problems in staff and visitors, whether work-related or not
- health promotion contributing to national intelligence on occupational ill-health by notifying an employer of a reportable disease, or notifying the Health and Safety Executive of a reportable accident in the surgery.

Roles of other members of the primary healthcare team in relation to the occupational health of the patients and of staff, and what tasks they could undertake

The practice nurse can have a key role in:

- liaison with occupational health services
- reviewing and reassessing chronic illness or disability in relation to fitness for work
- assessing psychosocial aspects of incapacity or premature retirement on the family.

The health visitor and community psychiatric nursing staff can have roles:

- in relation to the psychosocial aspects of work and health.

The practice manager has a role in:

- risk assessment in the practice
- development and implementation of health and safety policies.

All staff have responsibility for reducing risks from cross-infection by complying with waste disposal policies and procedures.

Occupational health services

Occupational health services exist to support an employer's commitment to health and safety, by advising management and employees, and promoting the highest standards of physical and psychological well-being. They only exist (in Great Britain) because of the commitment of employers in both public and private sectors. Great Britain and Northern Ireland still differ from the rest of the European Union (EU) in provision of workplace healthcare. Most other member states have legislation requiring the provision of occupational health services for many, if not all, workers.

The type and extent of professional occupational health services therefore varies greatly from industry to industry. The motives behind the development of occupational health services also vary.

Occupational health services have a responsibility to the employer whilst maintaining the confidentiality of the individual; they provide advice on statutory responsibilities in relation to health in the workplace and the environment, and on the management of absence attributed to sickness.

GPs may be contacted by an occupational health service for information – within the bounds of patient confidentiality – or because one of their patients needs specific help with their health in relation to their work. The provision of an occupational health service to patients and their employers is not (currently) part of General Medical Services, although GPs have a key role to play.

In 1992, the HSE commissioned a comprehensive survey of occupational health services.[5] This showed that there are approximately 1.1 million private sector establishments employing around 16 million people. The public sector employs a further 5.8 million. Of these firms, roughly 14% now employ a doctor or nurse, but this represents only 34% of the workforce. Twenty-eight percent of employees have no occupational health cover (not even First Aid).

There still remains much to be done to ensure the effective management of health and safety in the NHS, despite an increase in NHS occupational health nursing staff and consultant occupational physicians. Many large NHS Trusts now employ a consultant in Occupational Medicine, but smaller NHS units may rely on a nurse-led service, and employ a clinical assistant on a sessional basis. Occupational health services for general practitioners and their staff are still uncommon.

Specific legislation has improved awareness of hazards and risks in the workplace and has gone some way to improve the spread of occupational health and safety advice to industry. Major disasters have heightened health and safety awareness and led to investigations into standards. These include the Bradford City football stadium fire, the sinking of the *Herald of Free Enterprise* at Zeebrugge, the Kings Cross Underground fire and the Piper Alpha oil rig fire in which 167 men died. Public awareness of safety and 'safe systems' has been further heightened recently with several serious rail accidents in the London area. Safety issues have therefore predominated in recent years, but the increase in numbers of asbestos-related diseases, occupational asthma and occupationally acquired musculoskeletal and skin disorders has raised awareness of the need to identify and control occupational risks amongst both the public and employers. One specific Health and Safety Executive (HSE) initiative aimed at raising awareness was the 'Good Health is Good Business' campaign, which illustrated a number of workplace health risks, and described health risk management strategies for employers.[6]

An important development in occupational health is the work, led by the HSE, on an *Occupational Health Strategy for Great Britain*.[7] The strategy, which was published in summer 2000 is a joint commitment by the Government and others concerned about occupational health outside Government to work together to reach the following common goals:

- reduce ill-health both in workers and the public caused, or made worse, by work
- help people who have been ill, whether caused by work or not, to return to work
- improve work opportunities for people currently not in employment due to ill-health or disability and
- use the work environment to help people maintain or improve their health.

The strategy is being taken forward in five key programmes of work, one of which is to ensure that appropriate mechanisms are in place to deliver information, advice and other support on occupational health. An important plank of this programme will be the implementation of the recommendations of the Health and Safety Commission's Occupational Health Advisory Committee in their report on *Improving Access to Occupational Health Support*,[3] which was published at the same time as the *Occupational Health Strategy for Great Britain*.

In 1999 a partnership between the Department of Health and the Health and Safety Commission encouraged businesses to sign up to the *Healthy Workplace Initiative*. The first part of the 3-year strategic plan focused on raising the profile of occupational health with a key programme of preventing and managing musculoskeletal disorders.

The NHS is a relatively high-risk industry. Too many NHS employees are injured in the course of their work and too many patients die or are injured because a recognised hazard was not avoided or controlled. Given this picture it is surprising that so little attention has been given to occupational health services in the NHS.

The roles of occupational health professionals

The occupational physician

Occupational physicians are primarily involved in preventing ill-health arising at work, and promoting physical and psychological well-being of staff. They do perform personal consultations of a diagnostic and advisory nature, but then communicate with general practitioners and hospital specialists who are responsible for disease management. Occupational physicians also have an important role in advising management about accommodations and adjustments which may be needed for an employee to comply with the relevant legislation, particularly the Disability Discrimination Act 1995.

Whilst it is not strictly necessary to possess a higher qualification in occupational medicine to practise as an occupational physician, most employers now look for some evidence of training in the specialty. The Faculty of Occupational Medicine of the Royal College of Physicians, London, awards a Diploma (DOccMed) for the non-specialist doctor who has undertaken a course of study and passed an examination. Doctors pursuing a specialist career, or who wish to study occupational medicine in more depth, will take the examination for the Associateship of the Faculty (AFOM). Further information on these qualifications can be obtained from the Faculty.

The occupational health nurse

Many occupational health nurses are registered general nurses with a higher qualification in occupational health nursing. They are often called occupational health practitioners in industry. The Occupational Health Nursing Certificate (OHNC) and Diploma (OHND), have now been superseded by university degree courses – usually part-time – in occupational health. This level of training and experience enables the occupational health nurse to undertake a wide variety of tasks, looking after the day-to-day management of the occupational health service, providing immediate care, administering medications under standing orders from the occupational health physician, investigating cases of ill-health in the workplace, undertaking workplace environmental visits and so on. The nurse is likely to be the first point of contact for employees, managers and other health professionals.

The occupational hygienist

Some larger organisations may employ an occupational hygienist (often a chemist or engineer with specialist training) to assess the level of risk in the working environment and to give advice on reducing risk, or on improving the environment. The hygienist will be able to measure levels of fumes, noise, light, dust, etc., and will advise on methods of eliminating or reducing the problem. Specialist occupational hygienists with expertise in dealing with ionising radiations (health physicists) are employed in the nuclear industry.

The safety officer (or health and safety officer)

Many large organisations will have designated safety officers, trained to interpret health and safety legislation, identify hazards in the workplace and to give advice on general matters of safety. Their day-to-day work is largely concerned with accident reporting and prevention, risk assessment under the Control of Substances Hazardous to Health Regulations (COSHH)[8] and organisation of health and safety training. Their work will often overlap with that of an occupational hygienist, and sometimes with that of the occupational health nurse, and they will work closely with the training and human resources departments.

Safety officers come from a variety of backgrounds, often from a science or engineering profession relevant to their employment, and sometimes from the Health and Safety Executive. They will have undertaken formal study, leading to a recognised qualification in health and safety, and membership of a professional body such as the Institute of Occupational Safety and Health. Specialist safety officers address topics such as biohazards or radiation, and some general safety work is also carried out by environmental health officers.

Ethics and confidentiality

Any doctor or nurse dealing with occupational health issues must combine the need to maintain strict medical confidentiality in relation to an individual patient with the need to provide the employer with useful advice. This balance is still misunderstood by many employers, employees and doctors, who believe that a doctor or nurse may adopt a lower standard of medical ethics in relation to occupational health issues.

All communication between the occupational physician and management that relates to confidential medical information about a specific employee should be with the employee's informed and written consent. Medical details need not be given to managers, and advice on an individual's fitness for work can usually be given in general terms. An occupational physician or nurse communicating with a patient's employer should always try to make contact through the occupational health department (where one exists), to avoid the release of confidential material to personnel departments or other managers. It is also important to remember that any report you as a GP provide will have to comply with the requirements of the Access to Medical Reports Act (1988), which gives the patient rights of consent, access and comment. Reports that are provided to the Department of Social Security in relation to occupational health matters are not covered by the Medical Reports Act 1988; so patients do not have a right of access to see a report before it is returned to the Department. But patients do have a right to see any evidence which was used in the benefit decision-making process.

A GP has a vital advocacy role to play in relation to their patients' health at work. Many patients do not know how to use health services effectively and do not have a strong voice or representation at work. A well-informed GP may be able to offer the employee and employer specific advice (although they are not *required* to provide specific advice to employers in relation to occupational health issues). A GP may also be able to give an occupational physician or nurse an assessment of the wider family and social issues involved in a case of work-related ill-health from their knowledge of the patient's family – with their patient's written consent. Whilst advocacy is a part of an occupational physician's role – particularly where employees may not understand how to approach managers about health issues, the occupational physician is not the employee's personal medical attendant, and therefore must be careful to give advice that is as fair and objective to both parties as possible.

From time to time, occupational health professionals are asked to discuss issues with trade unions or other employee groups. Impartiality is vital, so that individual employees or their representatives clearly appreciate that the doctor or nurse is not allied with management or employee groups, and are giving objective and professional views.

The effects of work on health

The occupational history

If a doctor or nurse is not to miss important occupational factors in the aetiology of disease, or miss the significance of illness on fitness to work, a good occupational history is essential.

The proper questions are:

- What is your job?
- What do you do at work?
- Do you have any other jobs?
- What other jobs have you done?

A patient's job title may be meaningless. Without enquiring or without specialised knowledge, a doctor or nurse will simply not know what, for example, a 'health physics monitor' or 'crane banksman' does at work. It is important to know what tasks the patient actually performs, and particularly to what *hazards* he or she has been exposed, including chemicals, biological agents, dust, fumes, noise, heat, cold, radiations (ionising and non-ionising), and psychological or physical stresses.

It may be important to record every job that the patient has had – even if that is time-consuming. Occupationally caused cancers may have a long latent period between exposure to the causative agent and development of disease, and conditions such as noise-induced deafness may develop slowly, or have been caused in early working life. It may also be helpful to record hobbies and DIY, which can cause 'occupational' diseases.

Hazards in the workplace

For every workplace *hazard* – whether chemical, biological, physical or psychological – there is a degree of *risk*, which is the likelihood of the hazard causing harm in a particular situation. Most organs of the body can be harmed by adverse working conditions and many substances in the workplace have acute or chronic effects on health.

A primary task of occupational health practice is to predict that risk from a knowledge of the type of hazard, the working environment, and the degree of exposure (of individuals or groups) to the hazard. Having assessed the risk, occupational health and safety professionals advise employers on how to minimise it.

Health surveillance

'Health surveillance' may take many forms, ranging from simple health questionnaires through to regular medical examinations, with biological monitoring of blood or urine. Monitoring the health of workers is undertaken to detect early signs of work-related ill-health.

▼

Note all hazards in the workplace.

Employers must carry out statutory health surveillance on employees exposed to certain hazards, for example lead, ionising radiations, hazardous chemical or biological agents and asbestos. Health surveillance should also be conducted where an employee is exposed to a residual risk of harm from any workplace agent after control measures have been taken into account and where a number of criteria apply. This might include people working with respiratory sensitisers, such as laboratory animals or iso-cyanate paints, or in noisy environments. There must be valid ways to detect the work-related condition, and the surveillance must help to protect the worker from harm.

The type and frequency of the health surveillance will depend very much on the hazard and the magnitude of risk. Targeted health surveillance is always better than a 'routine medical' without a clear aim. The results of health surveillance undertaken on your patients in their workplaces should be available to you as their GP. If abnormal results are found, the occupational physician will usually seek the patient's consent to notify you as the GP so that you are fully informed and can participate in following up the medical management. Even where health surveillance is normal, the results will be made available to you, if you request them with your patient's written consent. For example, if a scrapyard worker presents in your surgery with recurrent abdominal pain, you will need to know that a recent blood lead test was within normal limits, to exclude lead poisoning.

The effects of health on fitness for work

The diagnostic and clinical management aspects of general practice and occupational medicine overlap, but the balance of an occupational physician's work tends more towards the management of ill people in the workplace or health promotion, rather than the diagnosis and management of occupational disease.

Common day-to-day issues for an occupational physician include:

*   pre-employment health assessment
*   general health promotion
*   sickness absence
*   rehabilitation/resettlement
*   ill-health retirement.

These areas overlap with the work of general practitioners, who are required under their NHS Terms of Service to provide advice to their patients about whether, as a result of a medical disease or disablement, the patient should refrain from their usual type of work. The advice is recorded on an official document (e.g. form Med 3) which may be accepted by an employer, or by the Department of Social Security, as medical evidence that the person is incapable of their regular occupation.

Pre-employment health assessment

It is naïvely thought by many employers that a 'pre-employment medical' will predict future illness and assist in the selection of a superfit workforce. This is a false assumption, and in many cases all that is really useful is a carefully designed health questionnaire, administered by a nurse with referral to a doctor if there are positive findings.

A GP may be asked to give information on a patient's medical history when that patient applies for a job. The familiar 'medical history form' used by insurance companies and some employers can lead to ethical problems. A prospective employee may be asked to give blanket consent for the employer to approach you for a medical report, and may feel that he or she will be discriminated against if they refuse consent. Any such medical information should go directly to the occupational health service if one exists rather than to a personnel department. If possible, it is helpful for the GP to discuss the content of a report with the patient to ensure that consent is 'fully informed'.

The Disability Discrimination Act 1995 (DDA) makes it illegal for employers to discriminate against people who have disabilities that make it difficult for them to carry out normal day-to-day activities. The disability must be substantial and have a long-term effect; that is, it must last more than 12 months. Employers of fewer than 15 people are excluded from this, but are encouraged to follow good practice when employing new staff. An employer must also make 'reasonable' adjustments to help a disabled person do a job. This may include physical adaptations to the workplace, changes in hours, etc. However, if an employee's health makes them incapable of doing the proposed job, or if the work might make their health worse or be unsafe, these considerations will override the provisions of the DDA.

Ethical, legal and technical issues involved in pre-employment or pre-placement health screening are complex.

Health promotion

Health promotion in the workplace has been encouraged by the government in recent years. A working population is 'captive' and often receptive, but any occupational health involvement in general health promotion must be carefully planned to avoid duplicating the health promotional work of a GP or giving a patient conflicting advice.

Examples of workplace health promotion which may cause problems for the primary healthcare team include cervical cytology and coronary risk factor screening. Although the practice of running cervical cytology sessions at work is becoming less common, female employees may appreciate the opportunity to attend such a session. The recording of attendance, quality control, reporting of results and follow-up of abnormalities are all areas which may cause the practice some concern. Similarly, the measurement of blood pressures and cholesterol levels in the workplace can lead to GPs and practice nurses being expected to counsel patients and follow up alleged abnormalities, without any say in how or why these tests are being performed.

Sickness absence

Sickness absence is a heavy burden on employers. Over the five years to 1991 the average number of days lost per year through industrial action was 2.8 million – less than a tenth of the number lost through work-related illness and injury. Table 1.1 shows days lost due to illness caused by work (or made worse by work), broken down by disease group.[2]

Table 1.1   Days off work by illness category in 1990 (England and Wales only)

Disease category	Days lost (000s)	Days lost per case
Stress/depression	1672	20
Headache and eyestrain	250	6
Upper respiratory disease	385	10
Lower respiratory disease	607	18
Asthma	302	14
Skin disease	132	10
Musculoskeletal problems	4319	18

Occupational health departments have a role in assisting managers to cope with absence attributed to sickness in staff. Multiple short absences are usually related to social or domestic problems rather than significant illness, and alcohol or other substance abuse may present in this way.

A worker with long-term sickness absence should be reviewed medically to assess whether the condition is likely to improve or to continue to interfere with the normal job, and whether rehabilitation or redeployment into a different job is possible.

Liaison between an occupational health service and the general practitioner over sickness absence is vital. A GP may be unaware of repeated short absences from work that their patients are attributing to sickness. The certifying GP may not wish to put a precise diagnosis on a sickness absence certificate for reasons of confidentiality. In these situations, an employer or occupational health department may request further information from the GP – but only with the patient's informed and written consent.

Rehabilitation, resettlement and retirement

Rehabilitation in occupational health terms simply means getting the worker back to work in a job which is appropriate to their medical condition – by temporary or permanent redeployment, or by specific redesign of the job. Again, communication between the employer or occupational health service and the GP will be of great help. The GP will have knowledge of specific medical factors which may limit a patient temporarily or permanently, and may also have background information on hospital investigations, treatment and re-habilitation after illness or injury.

Ill-health retirement may be necessary if the worker is permanently unable to do the job for which they were contracted and no suitable alternative can be found. Discussion between the occupational physician and the individual's GP is important before considering ill-health retirement, and careful thought must be given to the physical and psychological consequences – including the possible financial difficulties. It is important for both the GP and occupational physician to have a working knowledge of sickness and disability benefits available, to help the patient claim appropriately.

Permanent disability need not be a bar to future employment. Job Centres have specially trained Disability Employment Advisers (DEAs) who offer help in placing disabled people into appropriate jobs, and in recommending or providing individuals and firms with special aids or modifications.

- Retirement from their regular occupation on the grounds of ill-health or disability does not mean that the person cannot continue to work.
- There are a range of employment-related services available to GPs and their patients aimed at maximising work potential and giving people an opportunity to continue to make a positive contribution whenever possible (details are provided to all GPs in the IB204 Guide issued by the Department of Social Security[9]).
- State sickness benefits such as the Incapacity Benefit can often be a far less attractive option compared to more active intervention on the part of the GP, the patient and the occupational health services.

Specific problems

Workers with certain medical conditions such as epilepsy, diabetes and pregnancy need special consideration in terms of fitness to work. The two cases below illustrate how such conditions might be approached.

An 18-year-old student asks your advice on whether she should enter nurse training. She has had epilepsy since childhood, and is taking regular doses of sodium valproate. What clinical and occupational issues would you take into account in advising her?

There are wider implications than the simple clinical issues here. If the patient gets little warning of a fit, they may not be safe to handle patients, including children, or work in an operating theatre. Consider what a patient's reaction would be if a nurse collapsed whilst tending to them and the likely prejudices or anxieties of nurse managers, doctors and colleagues.

Any entrant to a nursing course will be required to undergo pre-employment health surveillance by an occupational health service. This is where the GP's knowledge of the patient and their medical problems can be used in conjunction with the occupational health department, covering all the points listed above, and giving the patient and her prospective trainers/employers advice on her suitability for a nursing career.

A woman who is 16 weeks pregnant consults you as she works in a local chemical factory and she is worried about exposure to chemicals.

Ask precisely what chemicals she works with or might be exposed to. If she is able to give you a list of chemicals, the names may not mean much to you, and you may need specific advice on whether the chemicals pose any threat to the pregnancy. Some chemicals may cause cancer, others may cause heritable genetic damage or risk to the unborn child or to breastfed babies. If the patient's employer has an occupational health service, they will have the information on the chemicals, or will have access to that information. The Health and Safety Executive may also be able to advise on specific chemical exposures.

The law requires employers to assess the risks to all employees including expectant and new mothers – that is, workers who are pregnant or who have given birth within the previous 6 months, or who are breastfeeding – and to do what is reasonably practicable to reduce the risks. Guidance on this is contained in an HSE publication *New and Expectant Mothers at Work – a Guide for Employers*.[10] You may be asked to confirm in writing that your patient is pregnant, but it is then the employer's responsibility to decide whether the

continued overleaf

risks to the mother or the unborn child are such that she needs to have her working hours or conditions changed, or if she should be redeployed into alternative work, or whether (if these are not possible) she should be given paid leave for as long as is necessary. If there is an occupational health service, you could check that these steps were being taken to help you reassure the patient.

Some jobs have particularly strict health requirements. These include commercial driving and flying, off-shore work or sea-faring and professional diving. Good advice to the patient and prospective employer is essential to avoid problems – a perforated gastric ulcer on a ship at sea is likely to be fatal.

Health and Safety: the legal framework

All modern British health and safety law stems from the Health and Safety at Work, etc. Act (HSAWA) 1974. This primary statute requires employers to protect the health, safety and welfare of their employees whilst at work, so far as is reasonably practicable. Employees also have responsibilities to protect their own health and that of their colleagues. Appendix 1 has extracts of the most important sections of the Act, and a summary of other important legislation.

One key requirement of the HSAWA is that all employers of more than five people must have a written statement of their arrangements for managing health and safety, a Health and Safety Policy. If you employ more than five staff (receptionists, nurses, secretaries, etc.), you should have made your own arrangements for managing health and safety; this area is explored in more detail in Chapter Seven.

A number of Regulations have been made under the Health and Safety at Work Act, covering such diverse areas as First Aid, Display Screen Equipment, Manual Handling Operations, and the Control of Substances Hazardous to Health. One of the most important sets is the Management of Health and Safety at Work Regulations (1999). These Regulations describe the statutory requirement for *risk assessment*, the need to put control measures in place, and the need to audit whether the controls are adequate. The rationale for health surveillance is also described in detail.

Health and safety legislation is enforced by the Health and Safety Executive, and sometimes by local authorities. Health and Safety Inspectors (previously called factory, agriculture or mines inspectors) and Environmental Health Departments will investigate cases of illness or injury caused by work, monitor compliance with statutory requirements, issue orders to change or cease certain dangerous activities ('Improvement' and 'Prohibition' Notices) and will prosecute individuals and firms if necessary.

The HSE obtains medical advice from the Employment Medical Advisory Service (EMAS), staffed regionally by qualified occupational physicians and nurses. EMAS can investigate clusters or single cases of occupational disease, and can give advice to GPs about suspected work-related health problems.

References

1   Health and Safety Executive (1999) *The Costs to Britain of Workplace Accidents and Work Related Ill-health in 1995/96* (2e). HSE Books, Sudbury.

2   Health and Safety Executive (1998) *Self-reported Work-related Illness in 1995: results from a household survey.* Health and Safety Executive, London.

3   Occupational Health Advisory Committee (2000) *Improving Access to Occupational Health Support.* HSE Books, Sudbury.

4   Health and Safety Executive (1993) *Your Patients and Their Work: an introduction to occupational health for family doctors.* HSE Books, Sudbury (currently out of print).

5   Bunt K (1994) *Occupational Health Provision at Work* (HSE Contract Research Report No 57). HSE Books, Sudbury.

6   Health and Safety Executive (1998) *Good Health is Good Business: an employer's guide.* HSE Books, Sudbury.

7   Health and Safety Commission (2000) *Occupational Health Strategy for Great Britain.* HSE Books, Sudbury.

8   Health and Safety Executive (1999) *Control of Substances Hazardous to Health Regulations 1999. Approved Codes of Practice L5.* HSE Books, Sudbury.

9   Department of Social Security (2000) *Guidance for Doctors. 1B204 Guide.* Department of Social Security, London.

10 Health and Safety Executive (1994) *New and Expectant Mothers at Work: a guide for employers,* HSG 122. HSE Books, Sudbury.

Reflection exercises

If you want to consolidate your learning, you need to spend some time thinking about how what you have read applies to your own practice. Why not do one, or all, of the following exercises?

Exercise 1. Think back over the past year, and apart from any occupational health practice of your own, note down and analyse instances of when you have been approached by a patient, an employer or an occupational health service, with a specific occupational health problem and comment on the reasons for that approach. If you want to complete more than two examples please photocopy this page. *This Exercise might take 1 hour.*

Audit of recent contacts with local occupational health services

Case 1:   Main reason for contact?

Who initiated the contact (patient, doctor, employer, etc.)?

How did you respond (by phone, letter, visit to workplace, ignored contact)?

Did you understand all the issues?

Were you happy with the contact and the interchange (was it ethical/did it have a satisfactory conclusion for your patient)?

Did you have the patient's written and informed consent to release information?

What might you do differently in future?

Case 2:   Main reason for contact?

Who initiated the contact (patient, doctor, employer, etc.)?

How did you respond (by phone, letter, visit to workplace, ignored contact)?

Did you understand all the issues?

Were you happy with the contact and the interchange (was it ethical/did it have a satisfactory conclusion for your patient)?

Did you have the patient's written and informed consent to release information?

What might you do differently in future?

Exercise 2. Review at least 30 sets of your patient records, selected at random (for instance using a random number system from your age–sex practice register), for patients aged between 16 and 65 years old. Indicate whether the current occupation and former occupations of each patient are recorded, and the likely accuracy and completeness of that information. Exactly how you undertake your review will depend on whether you keep mainly paper-based or electronic records – and if electronic, whether you have an appropriate search facility. You might delegate the collection of data to someone else in the practice. *This Exercise might take 1 hour.*

Audit of patient records to determine extent to which occupational history is recorded in medical notes

	Any occupation recorded? Yes/No	Present occupation recorded? Yes/No: if yes, what?	Completeness of occupational record? Yes/No
1			
2			
3			
4			
5			
6			
7			
8			
9			
10			
11			
12			
13			
14			
15			
16			
17			
18			
19			
20			
21			
22			

continued overleaf

23	
24	
25	
26	
27	
28	
29	
30	

Can you draw a conclusion from this audit as to whether you need to improve your recording of patients' occupational histories. Are all GPs and practice nurses equally careful about recording occupational histories?

Exercise 3. Use the information from the data gathered and your conclusions to draft guidelines for everyone in the practice as to future routine recording of patients' occupational histories. Indicate how the guidelines will be applied: who should be making the record and when, how it will be recorded, the detail of what should be recorded, how adherence to the guidelines will be monitored and how the recorded information should be used. Agree the draft guidelines with others in the practice after appropriate revisions; implement the guidelines. *This Exercise might take 2 hours.*

CHAPTER TWO

The relationship between work and health

Ill-health may arise because of exposure at work to *hazards*: physical, chemical, biological or psychological. Ill-health may also arise from *accidents* at work.

It is not always easy to determine whether an illness is genuinely 'occupational' or 'work-related'. There is a spectrum of disease from those where the relationship between exposure to specific causative factors and onset of illness is clear, to those where a possible causal relationship is weak or inconsistent.

Workplace hazards

The words *hazard* and *risk* are often used incorrectly as being synonymous. In occupational medicine, hazard is the inherent property of the substance or situation to cause harm, and risk is the likelihood of the harm occurring under any given circumstances.

You may never need to explore the nature of workplace hazards in depth, but you should be aware of simple classification into chemical, biological, physical and psychological.

- *Chemical* hazards include solid chemicals, liquids, gases and fumes.
- *Biological* hazards include bacteria, viruses, fungi and other micro-organisms, including the more exotic genetically modified organisms encountered in research laboratories.
- *Physical* hazards include noise, heat, light and unsafe workplaces or work equipment.
- *Psychological* hazards include all the common causes of stress in the workplace, from job overload through to bullying.

The risk to individuals from these hazards depends on 'how much' hazard is present, and on the circumstances and extent of the exposure. For example, the solvent 1,1,1-trichloroethane is found in a variety of places. Until recently, it was a common solvent in typist's correcting fluid, and was also used to degrease metal in manufacturing processes. Although the use of this substance is being phased out under the Montreal Protocol (as it depletes the ozone layer) the risk from the quantities of solvent in a small bottle of correcting fluid is rather different to the risk from a degreasing bath. This 'risk assessment' is the fundamental basis of health and safety practice in the workplace and is the basis of most of the modern legislation.

The Control of Substances Hazardous to Health Regulations (COSHH)[1] require 'suitable and sufficient' risk assessment for the use of all substances (chemical or biological) which may cause harm. It is not sufficient simply to list the properties of the substance, as in a 'hazard data sheet' or 'material safety data sheet'. It is necessary to estimate the level of risk from the substance, given the way the substance is handled in the workplace, and the level of exposure of individuals.

Stages in a risk assessment for the general practice team's exposure to such hazardous substances as blood and body fluids in their work

1   Identify the hazards (hepatitis B and C, human immunodeficiency virus (HIV), etc.) and how likely they are to cause infection in accidental exposures (by needlestick, etc.).
2   Describe under what circumstances you and your staff are exposed to the material (taking blood samples, testing urine specimens, etc.).
3   Assess how you can protect yourself, and reduce the potential exposure (by using enclosed venepuncture systems, vaccination against hepatitis B, wearing gloves and other protective gear).
4   Describe the overall risks from infection, having taken into account any special circumstances of your work (e.g. giving intravenous medication in a patient's home in poor light at 4 am!).

Some knowledge of the toxicology of the hazard may be necessary to undertake risk assessment properly. There are legal limits for workplace exposures to some chemical, physical and biological hazards; for example, there are limits set for exposure to airborne chemicals and to physical hazards such as noise. The 'exposure standards' of importance are:

*   occupational exposure standard (OES) – the concentration of an airborne substance, averaged over a reference period, at which, according to current knowledge, there is no evidence that it is likely to be injurious to employees if they are exposed by inhalation, day after day, to that concentration
*   maximum exposure limit (MEL) – the maximum concentration of an airborne substance, averaged over a reference period, to which employees may

be exposed by inhalation under any circumstances. It is not a safe limit and employers have to keep exposures as low as is reasonably practicable below the MEL.

An OES or MEL is usually expressed as p.p.m. or mg/m^2 of a substance over an 8-hour working day, or for short-term exposures, over a 15-minute reference period. The decision on whether limits are expressed in terms of 8 hours and/or 15 minutes is based on the nature of the toxicological hazard.

Essentially, the OES for a particular substance is set at a level at which there is no indication of risk to health; for an MEL, a residual risk may exist, and the level set takes into account socio-economic factors. For a substance to be given an OES, there must be a clearly identified critical health effect, and a 'no observed adverse effect level'. In reality, there may not be a clear dividing line between 'harm' and 'no harm' and the experimental data on which these levels are set may be incomplete or limited. Temporary excursions above the OES should not be expected to lead to serious health effects and compliance with the OES should be reasonably practicable. If it is not, then consideration is given to an MEL. For a substance to be given an MEL, there must be serious concern for the health of workers exposed to the substance. MELs are often assigned to carcinogens and respiratory sensitisers.

The HSE publishes an annual list of Occupational Exposure Standards (EH40)[2] that gives further details on the derivation and implication of exposure standards.

Target organs and toxicology

You need to think about how the agent can enter the body and which organs or body systems might be affected, to appreciate how occupational diseases might occur in relation to exposure to industrial agents.

Routes of entry for chemical exposures include:

* respiratory tract (inhalation)
* gastrointestinal system (ingestion)
* skin absorption.

Chemical agents may cause local effects such as skin or eye irritation, even if they do not enter the body.

The properties of a chemical hazard determine how it might enter the body. You should consider the physical and chemical properties of the hazard: is it a dust, gas or fume and likely to enter through inhalation; is it a liquid or solid that could be ingested by accident or through poor hand-washing; is it fat soluble and likely to enter through the skin?

Once inside the body, a toxic substance will produce distant effects on target organs. The most common organs to be affected are the lungs, liver, nervous system, bone marrow, kidneys and skin.

The lungs are at great risk as they are the first organ to encounter inhaled hazards. The liver may be involved in the metabolism of toxic substances and the kidney will encounter toxic substances or their metabolites in the process of excretion. Many industrial chemicals are fat-soluble and so find their way to brain, spinal cord and bone marrow. Their toxic action may be prolonged by storage in body fat.

Individuals differ in their susceptibility to toxic effects of chemicals. Apart from the actual current level of exposure, the personal or medical factors that will influence the likelihood or severity of a toxic effect in an individual include:

- age
- sex
- genetic factors, e.g. atopy, enzyme deficiencies
- ethnic background
- coexisting diseases
- nutritional status
- fatigue
- coexisting exposures to other chemicals
- previous exposure to the same chemical.

Examples of occupational exposures and related diseases

The following list of occupationally acquired or associated diseases and the possible causes is not exhaustive, but indicates the variety of problems which may be encountered. Some are very rare, some very common, and whilst reading through the list, you might like to try to relate the condition and its likely cause to routes of entry into the body.

Condition	Jobs at risk	Possible causative agents
Respiratory diseases		
Asthma	Car paint spraying	Isocyanates
	Laboratory work	Small animals
	Joinery	Hard woods
	Electronics, soldering	Colophony
	Bakery, farming	Grain, flour improver
	Healthcare workers	Glutaraldehyde
	Heavy manual work	Exercise

continued opposite

Condition	Jobs at risk	Possible causative agents
Allergic alveolitis	Farming	Mouldy hay
Byssinosis	Cotton milling	Cotton dust
Pneumoconiosis	Coal mining	Coal dust
Silicosis	Mining, stone masonry	Quartz
Asbestosis, lung cancer and mesothelioma	Plumbers, electricians and builders	Asbestos
Rhinitis	Healthcare workers	Glutaraldehyde
	Laboratory work	Animal allergy
Skin diseases		
Contact dermatitis	Building trades	Cement
(irritant)	Painting	Polymers, solvents
	Laggers	Glass fibre
	Domestic work	Detergents
	Hairdressers	Chemicals, shampoos
Contact dermatitis	Wearing rubber gloves	Latex protein
(allergic)	Plastics, joinery	Epoxy resins
	Horticulture	Primula, etc.
	Hairdressers	Chemicals, etc.
Skin ulcers	Plating	Chrome
Acne	Metal work	Mineral oils
Epithelioma	Farmers	Sunlight
	Road-layers	Tar
Neurological diseases		
Peripheral neuropathy	Battery recycling	Lead
	Solvents	Methylbutyl ketone
Parkinsonism	Fire fighting	Carbon monoxide
Urogenital diseases		
Bladder cancer	Dye manufacture	Aromatic amines
Blood disorders		
Aplastic anaemia and acute leukaemia	Petroleum industry	Benzene
Cardiovascular disease		
Raynaud's phenomenon	Use of hand tools	Vibration
Infectious diseases		
Hepatitis B	Healthcare workers	Blood/body fluids
Leptospirosis	Farmers, sewage workers (different strains of *Leptospira*)	Rat/cattle urine
Ears, eyes		
Hearing loss	Heavy industry	Noise
Keratoconjunctivitis	Welding	Ultraviolet radiation
Corneal damage	Grinding	Chemicals, foreign body, splashes
Cataracts	Steel and glass workers	Infra-red radiation

continued overleaf

Condition	Jobs at risk	Possible causative agents
Musculoskeletal problems		
Tenosynovitis	Keyboard use	Repetitive movements
Epicondylitis	Plastering	Repetitive movements
Low back pain	Labouring	Heavy lifting and twisting
Psychological problems		
Stress	All occupations	Work overload, job insecurity, etc.
Anxiety/depression	Most occupations	As above

You may never see a case of leptospirosis due to exposure to rat urine in general practice, but it is one of the differential diagnoses for a pyrexia of unknown origin in a sewage worker. You will almost certainly never come across a case of aplastic anaemia from benzene exposure but there are many 'common' occupationally acquired disorders. Four of the most common problems (asthma, dermatitis, upper limb disorder and 'stress') will be explored in more detail in Chapter Six.

Examples of occupational health disorders

Five examples are included; tennis elbow, back pain and latex allergy are common whilst the other two – lead poisoning and farmer's lung are uncommon, but described here because they illustrate some general points about toxicology and pathology.

'Tennis elbow'

Lateral epicondylitis can arise spontaneously, but is common in workers who put strain on their forearm muscles by using hand tools or by working with their forearm pronated and their elbow partly bent. The problem is common in plasterers, joiners and bricklayers. It also occurs in computer operators who use a 'mouse' pointing device for much of their work, because of the need to grip the 'mouse' with the forearm pronated, and to move the mouse around, putting a strain on the insertion of the tendons into the lateral epicondyle.

Find out when the symptoms started and whether the patient thinks they are made worse by work. Ask about hobbies and sporting activities. Epicondylitis may not be work-related in a tennis player!

The options depend on the opportunities for alternative work, the availability of treatment and the severity of the problem. It may be unrealistic to ask a self-employed joiner to take several weeks off work whilst waiting for physiotherapy, but some patients will be able to avoid the tasks which exacerbate their symptoms. There is some evidence that early steroid injection improves the rate of recovery.[3] Discussion with the patient's occupational health department (if there is one) would be very helpful in deciding the best treatment options.

Low back pain

Low back pain (LBP) is one of the most common disorders that you come across in practice. Most adults (60–80%) experience LBP at some time, and it is often persistent or recurrent. Workers whose jobs involve heavy manual handling or repeated bending tend to report more low back symptoms, but most people in lighter jobs also have similar symptoms. The relationship between job demands and symptoms is inconsistent, and although people with heavy physical jobs may lose more time from work due to back pain, this may be because they have difficulty working with the pain, rather than because their job caused the pain.

For any worker who develops acute low back pain, the ways in which the GP manages both the clinical care and the occupational issues are extremely important. If you are satisfied that there is no serious spinal pathology[4] then dealing with the psychosocial issues and the attitude to returning to work becomes as important as managing the pain and immobility.

Communication, co-operation and identification of agreed goals between the worker, the GP, the worker's occupational health department (if there is one), and the worker's managers is fundamental. Most workers can continue working with LBP, or can return to work within a few days or a couple of weeks, if necessary to modified duties. There is no need to wait until they become pain-free.

The longer the worker is off work with LBP, the lower their chances of ever returning. Once a worker is off work for 4–12 weeks, they have a 10–40% chance of still being off work at 1 year. It is therefore extremely important to consider active rehabilitation before your patient has been off work for more than a few weeks. Changing the focus from purely symptomatic treatment to active rehabilitation reduces long-term disability. Rehabilitation programmes are not always easy to access, but referral for assessment and rehabilitation should be the norm for LBP continuing for more than about 6 weeks. Dealing with inappropriate beliefs (that pain is harmful), fear-avoidance leading to inactivity, or associated low mood and withdrawal are an integral part of rehabilitation, as are progressive exercises.

If the patient with LBP works in a heavy physical job, it may be necessary to explore alternative or modified duties to help them return to work. This should be combined with active encouragement to deal with the psychosocial barriers to returning to work, and help with learning to manage the pain. For workers in 'light' physical jobs, early return to normal work is essential, and there is no indication to advise people to refrain from work until they are pain-free.

Further detailed advice is contained in the Faculty of Occupational Medicine's book *Occupational Health Guidelines for the Management of Low Back Pain at Work: Evidence Review and Recommendations*.[5]

Lead poisoning

Lead poisoning is relatively rare, but is still possible in certain occupations. You might still come across a case. Those at risk include scrap dealers who handle sheets of lead and who break up car batteries for the lead, workers in the smelting and refining industries, people involved in work with lead paints or colours, demolition workers who handle or burn metal covered in lead-based paints, and workers in certain mixing and melting operations in the glass industry or pottery glazing. Some unusual occupations are also at risk, including stained glass workers, firearms instructors and even the cleaners in firing ranges! Workers with significant exposure to lead should be under statutory health surveillance (*see* below) by an Appointed Doctor or Employment Medical Adviser, and have regular blood lead measurements. But the sort of jobs that are at risk tend to attract temporary workers whose employers may not bother to notify the HSE of their work with lead.

There are two main routes of entry of lead into the body from occupational exposure: inhalation of fumes if lead or lead compounds are heated, and ingestion from poor hygiene and hand washing after handling lead.

The early symptoms of lead poisoning are: tiredness, anorexia, abdominal cramps, muscle pains and headaches. Anaemia occurs early in poisoning, because inorganic lead interferes with haem synthesis, causing a microcytic iron-deficient picture. In the late stages of poisoning, peripheral neuropathy is possible, classically causing wrist drop. Other signs, such as a blue line in the gums around carious teeth, are rare.

Diagnosis is made by looking for a raised blood lead level and a raised erythrocyte zinc protoporphyrin level. Lead blocks the conjugation of protoporphyrin and iron to form haem in the red cell, so protoporphyrin joins with the next most common metal in the body, zinc. The level of this abnormal metalloprotein is therefore the basis of biological monitoring of the effects of lead.

The treatment for lead poisoning is usually removal from the source of the contamination; it is highly unlikely that treatment with chelating agents will be necessary.

Legal aspects: all work involving exposure to lead is subject to the Control of Lead at Work Regulations 1998 and their supporting Health and Safety Commission Approved Code of Practice (ACOP). To avoid the possibility of lead poisoning, the ACOP specifies concentrations of lead in blood known as 'action levels' at which the employer must investigate and reduce exposures, and 'suspension levels' at which a worker should be removed from further work involving exposure to lead. The suspension levels are 60 µg/dl for adult workers and 30 µg/dl for women of childbearing potential; and 50 µg/dl for young persons 16–17 years old.

Lead is associated with high rates of spontaneous abortion, stillbirth and infertility, and there is an association between low-level environmental lead exposure *in utero* and mild intellectual impairment in childhood. An employer has a duty to inform workers of risks to health and if a woman knows that she is pregnant, she must immediately stop working with lead (*see New and Expectant Mothers at Work: a guide for employers*).[6]

Farmer's lung

This is an uncommon disorder these days, but one worth knowing about, particularly if your practice population covers a farming community.

Farmer's lung is one of the causes of extrinsic allergic alveolitis, an acute type III immunological reaction. Extrinsic allergic alveolitis describes a group of disorders caused by the inhalation of organic dusts. The causative agents are moulds, fungal spores or heterologous proteins. In the case of farmer's lung, the cause is *Faenia rectivirgula* (previously called *Micropolyspora faeni*). There is a seasonal variation in the incidence of farmer's lung, the peak occurring in late winter when farmers start to handle hay stored from the previous summer. There is usually a period of asymptomatic exposure over several years before symptoms start.

The classic picture is the development of flu-like symptoms, fever, malaise and joint pains, followed by cough and breathlessness, 3–4 hours after exposure. The symptoms resolve in 48 hours. There is usually loss of weight during an episode, but in individuals where exposure is low or sensitivity slight there may just be an insidious onset of breathlessness.

Diagnosis can usually be made from the occupational history, and respiratory function tests will show a restrictive pattern. Specific circulating antibodies can also be found.

The prognosis is variable with some farmers developing progressive disease which stops them working with hay, but most can continue to work if they wear respiratory protection.

Similar occupational disorders are:

- bird fancier's lung (due to inhaled pigeon bloom)
- mushroom picker's lung
- cheese washer's lung
- and the really exotic – paprika splitter's lung (from mouldy red peppers).

None of these are found very commonly in the UK but if you do not take an occupational history when confronted by a breathless Hungarian, you will miss the one remaining paprika splitter!

Latex allergy

The cause of latex allergy is rubber protein, found in cheaper rubber gloves made from poorly washed latex. The problem with latex allergy is that it can lead to severe allergic reactions in sensitised individuals. Cases have been reported of anaphylaxis in latex sensitive patients when operated on by a gloved surgeon.

Latex allergy most commonly presents with contact urticaria of the hands. Respiratory symptoms are not uncommon from airborne protein. In sensitised individuals, anaphylaxis can occur if latex protein is injected or introduced into the body, either as a gloved hand in surgical procedures, or even on a needle drawn through the rubber bung of a vial.

The diagnosis is largely clinical, but can be confirmed by tests for specific immunoglobulin E (IgE) or (with care) by skin-prick tests.

The treatment is simply avoidance which is not easy if the affected person is a laboratory worker or a surgeon. 'Hypoallergenic' gloves made from well-washed latex are available and alternative materials (nitrile, vinyl) can be used.

Accidents at work

Accidents are a major source of mortality and morbidity at work. Table 2.1 shows the number of injuries to employees, self-employed and members of the public for 1990–91 to 1998–99[7] in Great Britain.

Table 2.1 Number of injuries to employees, self-employed and members of the public

	1990–91	1991–92	1992–93	1993–94	1994–95	1995–96	1996–97	1997–98	1998–99
Fatal	572	473	452	403	376	344	654	667	625
Major[a]	31 203	29 707	28 722	29 531	30 996	30 968	65 014	58 615	52 409
O3D[b]	162 888	154 338	143 283	137 459	142 218	132 976	129 568	135 773	131 191
Total	194 663	184 518	172 457	167 393	173 590	164 288	195 236	195 055	184 225

[a]Major injuries are mainly fractures, serious eye injuries or accidents requiring immediate medical attention.
[b]O3D accidents are those which result in more than three days off work.

It is important to note that the figures from 1996–97 are not comparable with previous figures because of the introduction of the accident reporting system 'RIDDOR 95'. In particular, fatal injuries to members of the public including those resulting from acts of trespass or suicide on railways became reportable. Similarly, the change in definition of non-fatal injury to members of the public caused an apparent increase in these categories.

Generally, a reduction in figures does not mean that industry is getting safer. The figures are affected by the numbers of people employed in hazardous sectors. Part of the decrease in the fatal injury rate is accounted for by changes in patterns of employment in the mid 1980s. Between 1986/87 and 1994/95 there was a 12% increase in the number of employees in the low-risk service sector, and a reduction in the number of employees in the more hazardous sectors: energy (down by 49%), manufacturing (down by 19%), construction (down by 16%), and agriculture (down by 11%).

It is also worth bearing in mind that accidents often have immediate effects on health, whereas other workplace hazards may only cause ill-health after long cumulative exposures, or after a long latent period.

The single most common cause of fatal injury to employees in 1994–95 was fall from a height (23%). A further 37% were caused by being struck by a moving vehicle (20%) or a moving/falling object. Slips, trips and falls on the same level account for one-third of all major accidents but only 1% of fatalities in 1994–95. In the service sector, slips, trips and falls accounted for nearly 60% of all major injuries in 1994–95. There were no fatalities.

The fatal injury rate to the self-employed is highest in agriculture (12.9 deaths per 100 000 in 1994–95), followed by the construction and manufacturing sectors (3.2 deaths per 100 000).

Beneficial effects of work

It is too easy to assume that all effects of work on health must be negative; having a job has major psychosocial benefits with secondary benefits on physical health. Unemployment is associated with psychological morbidity (low self-esteem, chronic depression) and employment provides the financial basis for good nutrition and hygiene. Physically demanding jobs help to maintain muscular fitness and stamina, together with general cardiorespiratory fitness.

In general, an employer's health and safety policy and the management systems that are in place to implement and audit compliance with the policy, contribute to the health and safety of the workforce. In addition, an employer may introduce:

- no-smoking policies at work
- subsidised canteen food and 'healthy eating' options
- on-site (or subsidised) exercise or gym facilities, with inducements and encouragement for participants
- general health screening – blood pressure, cholesterol, etc.
- employment policies which encourage individuals to control their work and reduce stress.

▼

Observe the beneficial effects of work.

Occupational health services have a direct role in health promotion. They may initiate or support general health promotion (diet, smoking, exercise) and may be involved in specific projects, related more to the hazards encountered at work. Education of the workforce on matters of health, hygiene and accident prevention is important.

The involvement of occupational health services in general health screening is more controversial. Is it appropriate for occupational health services to undertake screening for coronary risk factors (smoking habits, blood pressure, cholesterol, diabetes, etc.)? Is it appropriate for them to undertake cervical cytology in the workplace?

If an occupational health doctor or nurse finds a coronary risk factor on 'health screening' (e.g. a raised cholesterol) then counselling the individual and communicating the results to the GP are vital, but it is still the GP who has responsibility for following up an abnormal result and for prescribing. General practitioners may have concerns about occupational health services undertaking these tasks unless they are satisfied about the quality of the service, the confidentiality of the results, and the communication and follow-up of abnormalities.

Demands of work: disability, chronic ill-health, rehabilitation and resettlement and ill-health retirement

Disability and chronic ill-health

The Disability Discrimination Act (1995) makes it illegal for employers of 15 or more staff to discriminate against people with disabilities when selecting, training or promoting employees. It requires employers to make reasonable changes to the workplace to accommodate people with disabilities and help them to work. In this context, disability is defined as any physical, sensory or mental condition which makes it difficult for the person to carry out normal day-to-day activities and is likely to last more than 12 months. It will also include conditions that presently have only a slight effect on day-to-day activities but which are expected to become more substantial, for example multiple sclerosis.

Contrast these two situations and consider whether ill-health retirement might be recommended for:

(a)   a 28-year-old secretary with neck pain following a whip-lash injury
(b)   a 57-year-old postman with osteoarthritis of both hips.

In case (a) it should be possible to modify the secretary's job to avoid exacerbating her neck problems, but in case (b) the postman is likely to be permanently unfit to continue work as a postman, unless redeployment into a sedentary job is possible.

Ill-health retirement may be necessary in some situations, where illness or injury are so significant that redeployment and resettlement are impossible. An employer is entitled to dismiss an employee on the grounds of ill-health, and an employee who is a member of a pension scheme is entitled to ask for early retirement with full pension benefits. Pension fund trustees usually require some sort of medical certification that the person is 'unfit', but the exact rules vary from scheme to scheme.

Some pension schemes simply require the employee to be unfit to do their usual job or any comparable job. Some pension schemes are more prescriptive, and may require certification of permanent incapacity to do any sort of work.

The Personal Capability Assessment (PCA) introduced in April 2000 focuses on residual capabilities rather than just 'incapacity'. The postman may well be able to undertake another form of work that is less physically exacting. Medical advice to inform benefit decision making is provided by a doctor 'approved' by the Secretary of State for Social Security and not the GP or the occupational physician.

In assessing whether changes they could make to the workplace or to the way the work is done are 'reasonable', the employer is allowed to take into account how much the changes would cost and how much they would help. The Act does not apply to operational staff employed in the armed forces, the police, the prison services, the fire services or to anyone employed on board ships, hovercraft or aeroplanes.

Employment legislation prior to the Disability Discrimination Act (DDA) required employers of more than 20 people to have a quota of 3% disabled, a quota that was frequently not filled.

Many patients with disabilities did not register as disabled under the 'old' system. In terms of job prospects, there are advantages and disadvantages to an individual of admitting that they have a 'disability'.

Advantages. Up to the introduction of the Disability Discrimination legislation, there were two occupations reserved for those registered disabled: car park attendant and electric lift attendant. Positive discrimination for other jobs, by enlightened employers (or employers wishing to fulfil their quota of 3% disabled under the old scheme) was possible.

Disadvantages. Negative discrimination by many employers who still do not attempt to accommodate disabled employees, leads many people with 'disabilities' to try to make light of them or deny them.

The DDA legislation is aimed at giving disabled people a fair chance of employment in all jobs and should encourage applicants to acknowledge their physical or psychological limitations in the knowledge that they will be treated fairly.

Employment *helps* a disabled person, or one with a chronic illness through the general benefits of work, together with improved morale and self-image for those able to be productive in paid employment despite a disability.

Contrast these two situations and consider whether ill-health retirement might be recommended for:

(a) *a 52-year-old chemical process worker with a long history of asthma, on maximum doses of inhaled medication, and whose asthma is exacerbated by exposure to chemical fumes*

(b) *a 35-year-old forestry worker who loses a leg in an accident at work.*

Redeployment and retraining should be possible in both cases, even if the employees are unable to return to their old jobs or old employers, so the pension fund rules may preclude them from drawing their pension.

If you are asked for your opinion on ill-health retirement, either directly by Pension Fund Trustees or indirectly by an occupational physician, it might be necessary to ask about the rules in detail before commenting on your patient's long-term prognosis. If the rules are very strict, the patient may find themselves without a job, and without an ill-health pension.

Rehabilitation and resettlement

An appreciation of the role of occupational health services and other agencies in resettlement and rehabilitation is very important for the general practitioner and practice nurse. It is not uncommon for a patient with a chronic medical problem to believe that they cannot, or should not, return to work, and for their GP to issue sick notes without considering the options for rehabilitation and resettlement. For a patient claiming a state incapacity benefit the GP will not have to issue further statements following the medical assessment by the DSS Benefits Agency. This will usually be at around 6 months into a spell of incapacity but is increasingly being applied earlier.

A GP has access to a number of agencies to help to get their patients back to work after illness or injury, including the employer's occupational health service, where there is one. Variables such as hours of work, physical and psychological demands and workstation design are critical. Advice on these matters (and the pace at which someone should be brought back up to 'full speed' at work) can be given to management by an occupational health service with additional advice from other services.

Advice from occupational therapy departments may be helpful in considering physical adaptations to workstations or aids to mobility.

Every Job Centre has a Disability Employment Adviser (DEA), who can give advice on all aspects of employment, and has contacts with employers and training agencies. The service of the DEA is open to those who are in employment as well as to the unemployed. For the under-18 year olds, the local Careers Service deals with questions of employment and disability.

A 42-year-old hospital porter suffers an episode of back pain with sciatica. The GP advised a short period of rest, followed by mobilisation and physiotherapy. After a month the porter's back pain had not entirely settled, and he went back to his GP for another sick note (Form Med 3). The GP might liaise with the occupational health service or DEA as an alternative to issuing another sickness certificate.

Occupational physicians know that the pressures on a GP's time are great. It isn't easy or always feasible to negotiate resettlement and temporary (or permanent) redeployment for your patients in the time you have available. Issuing a further Med 3 to a patient who says 'I don't think I'm ready to go back to that job, doc' is a reasonable response. Naturally, the patient's *attitude* to returning to work is vital, but if there is an occupational health service to assist with assessment of the workplace and work demands in relation to residual disability – and/or referral to the Regional Disability Service is feasible – then getting the patient back to work quickly becomes a real possibility.

In general, a patient may not be well served in the longer term by medical advice to refrain from work if more appropriate clinical management would allow them to stay in work or return to work as quickly as is reasonably practicable. In this case when providing advice to a patient about fitness for work the GP may wish to consider the following factors:

- the nature of the patient's medical condition and how long the condition is expected to last
- the functional limitations which result from the patient's condition, particularly in relation to the type of tasks they actually perform at work
- any reasonable adjustments which might enable the patient to continue working

- any appropriate clinical guidelines[4,5]
- clinical management of the condition which is in the patient's best interest regarding work fitness
- how to manage the patient's expectations in relation to their ability to continue working.

Employers, occupational health services and the DEA can refer individuals for further advice to the Regional Disability Service. These specialist teams can arrange assessment of physical or psychological problems. For example, one common reason for referral is chronic back pain. They assess the client's aptitude for – and attitude towards – certain kinds of work, and counselling by occupational psychologists can help appropriate redeployment at work. Further training and job rehearsal can be arranged. The Regional Disability Service (like occupational therapists) has access to a wide range of aids which may facilitate return to work.

The sort of aids and modifications that might help a secretary with severe rheumatoid arthritis to stay at work are:

- modifications to visual display units (VDU) and workstations
- special keyboards
- modified pointing devices
- voice activated word-processing systems
- special seating or adaptations for wheelchairs
- aids to assist handling paper and files
- help with transport to work
- adaptations to cars.

Other aids are available, including large screens, magnifying equipment, etc. Communication aids for the deaf and deafened, and help with transport (adaptations to a car or help with costs of taxi fares for employees who cannot use public transport) can be arranged.

Under the Access to Work scheme, costs of adaptations to the work environment, including the introduction of new equipment, can be covered (or partially covered) by grants from the Regional Disability Service or DEA. For example, for employees with poor visual acuity, magnifying screens or large screens can be provided for VDU work at a reduced cost to the employer.

Compensation and benefits

Knowledge of the basic rules of compensation and benefits for occupational diseases is useful in general practice. Patients are often unaware of their rights, and although you can direct the patient to the Department of Social Security, Citizens' Advice Bureaux, Trade Union or solicitor, it is worth knowing something of the issues involved. The Incapacity Benefit and Income Support (Disability Premium) are the state incapacity benefits most likely to be claimed by people of working age who are too ill to work. These benefits are claimed by more people and are worth more financially than Industrial Injuries Disablement Benefit.

The Industrial Injuries Disablement Benefit Scheme covers employed earners (not the self-employed). It excludes NHS employees who are covered by the NHS Injury Benefits Scheme, which is more generous. Workers can claim if they either:

- suffer a personal injury caused by an accident arising out of or in the course of their employment
- suffer from a 'prescribed disease', that is one designated by the Secretary of State, after consultation with the Industrial Injury Advisory Council; have a special risk for particular occupations, e.g. hepatitis B for those handling blood products; or carpal tunnel syndrome for workers handling vibrating tools.

Diseases are only prescribed if the occupational cause is well established and if the terms of the prescription can be framed to ensure that most cases falling under it will be occupational in origin.

Disablement benefit is a tax-free pension which continues for as long as the disablement lasts, regardless of whether the claimant is employed, or whether his earnings are reduced. To receive a pension, the claimant must be assessed as being at least 14% disabled, except for those suffering from pneumoconiosis, diffuse mesothelioma or byssinosis, where disablement must be at least 1%, and for occupational deafness, where disability must amount to 20%. The amount of pension depends on the degree of disablement; total loss of hearing or amputation of both hands can amount to 100% disablement, and it is not necessary for the worker to be bedridden. Assessment is by lay Decision Makers acting on behalf of the Secretary of State, informed by medical advice from a doctor specially trained and approved to assess such cases. A 'tariff' describes the extent of disability; for example, loss of a whole index finger = 14%, loss of a thumb = 30%. The two most common claims are for noise-induced hearing loss and vibration white finger (hand–arm vibration syndrome).

In addition to disablement benefit, the patient may be entitled to other benefits such as Disability Living Allowance, and reference should be made to a local Benefits Agency office.

Claims for benefit for accidents are more difficult to assess, and the claimant must show personal injury that has been caused by an accident arising out of and in the course of employment. Accidents are sudden events, and this might include an acute back injury, or contracting HIV from a needle-stick injury.

Civil claims for damages

A patient who suffers illness or injury can pursue a civil claim against his employers through the courts. To be successful, he must prove that his employer was *negligent* and/or that the employer was *in breach of statutory duty of care.* Negligence is a failure to take reasonable care which causes foreseeable damage. The employer has a duty to provide safe plant, safe personnel and a safe system of work, and what is reasonable will depend on individual circumstances. In the case of *Paris vs Stepney Borough Council* [1951], a worker who was known to be blind in one eye was not given eye protection and suffered total blindness after a penetrating eye injury. Although the risk of eye injury was small, if it did occur to this worker, it would be catastrophic and a higher than normal duty of care existed.

The case law surrounding negligence claims is complex, fascinating and quite alarming for employers – like GPs – who have a statutory duty of care! Further details can be found in *Occupational Health Law.*[8]

References

1   Health and Safety Executive (1999) *Control of Substances Hazardous to Health Regulations 1999. Approved Codes of Practice L5.* HSE Books, Sudbury.

2   Health and Safety Executive (2000) *Occupational Exposure Limits.* EH40. HSE Books, Sudbury.

3   Hay EM, Paterson SM, Lewis M, Hosie G, Croft P (1999) Pragmatic randomised controlled trial of local corticosteroid injection and naproxen for treatment of lateral epicondylitis of elbow in primary care. *BMJ.* **319**: 964–8.

4   Royal College of General Practitioners (1999) *Clinical Guidelines for the Management of Acute Low Back Pain.* Royal College of General Practitioners, London.

5   Faculty of Occupational Medicine (2000) *Occupational Health Guidelines for the Management of Low Back Pain at Work.* Faculty of Occupational Medicine, London.

6   Health and Safety Executive (1994) *New and Expectant Mothers at Work: a guide for employers.* HSG 122. HSE Books, Sudbury.

7   Health and Safety Executive (2000) *Health and Safety Statistics 1998/99.* HSE Books, Sudbury.

8   Kloss D (1998) *Occupational Health Law* (3e). Blackwell Science, Oxford.

Reflection exercises

If you want to consolidate your learning, you need to spend some time think-ing about how what you have read applies to your own practice. Why not do one, or both, of the following exercises?

Exercise 1. Work through these two case histories describing how you would approach these two conditions using what you have learnt from reading the material in this chapter.

1 Ann, the hairdresser

An 18-year-old trainee hairdresser consults you because of an eczematous rash on her hands and wrists.

(a)   *What are the possible causes?*

(b)   *What investigations should be undertaken?*

(c)   *What are the options for eliminating or managing the problem?*

After being investigated and treated, she tells you that she has a part-time job in a pub, working behind the bar and serving meals.

(d)   *What are the occupational health implications, and what advice would you give her?*

Three days after this consultation, you receive a phone call from the landlord of the pub in which the girl works. He wants to know what exactly is wrong with her, and whether she is 'safe' to work in the kitchen.

(e)   *How would you reply?*

This part of Exercise 1 might take 1 hour.

2 Rob, the demolition worker

A 38-year-old man presents in your surgery one Monday morning, com-plaining that he is 'tired all the time'. He says he has been off his food recently, and has had vague abdominal pains. He hasn't been to see you for several years; the last entry in his notes related to a hernia repair 12 years ago. He looks pale, and you suspect he may be clinically anaemic. You ask his occu-pation, and he tells you he is a demolition worker.

(a)   *What general medical and further occupational questions would you ask?*

(b)   *What investigations would you consider at this first consultation?*

He returns a week later, for the results of his tests. He has a hypochromic microcytic anaemia, and no evidence of occult blood in his stool.

(c)   *What is the differential diagnosis at this stage?*

You ask him more about his job, and he tells you that he is dismantling an old steel railway bridge, using oxyacetylene cutting tools. He felt quite well until this part of his job started, about a month ago. You suspect he may have been exposed to lead fumes in his job.

(d)   *Who would you speak to about this problem, and why?*

This part of Exercise 1 might take 1 hour.

Exercise 2. Visit two different workplaces where many of your patients work. Contact their employing organisation or the occupational health unit and ask if you can visit. Look at what several different workers actually do – and spot the potential hazards involved in their jobs. Make a list of the potential hazards for each of the jobs and list the physical and mental effects. For instance there may be hazards from lifting, obstacles on the floor, noise, chemicals, etc. for someone working in a particular factory with the possible effects of muscular pain, fractures from falls, deafness and skin conditions.

(i)   Describe the job: what exactly does it involve?
(ii)   List the potential hazards.
(iii)   Suggest possible effects linked to each hazard.

This Exercise might take you 6 hours.

CHAPTER THREE

Investigating occupational health problems

The clinical approach

Successfully identifying occupational disease depends on *asking for information* in an appropriate way. You may not be able to talk to groups of workers but you will be able to talk in some detail to the individual who believes he or she may have a work-related problem.

The occupational history

Taking a full occupational history is not always possible or necessary in every general practice situation. But where there is any suspicion that symptoms of disease may be work related – or that fitness for work may be compromised – a good occupational history is vital.

Questions that comprise an occupational history are:

- What is your job?
- What other jobs have you had in the past?
- What do you *do* at work?
- What hazards (chemical, physical, biological, psychological) are you exposed to at work?
- What are your hours of work/shifts?
- Do you have any other jobs or hobbies that bring you into contact with hazards?
- Has your job changed at all recently?
- Do your symptoms change (better or worse) during the course of the working day, or at weekends, or when you are on holiday?

- What do *you* think is causing your symptoms?
- Does anyone else at work have the same problem?

A patient's job title may not be very useful on its own. A description of what the patient actually does is usually interesting and may need a great deal of explanation. Because of the long latency of some occupational diseases, it may be necessary to take a full history of every job held since leaving school.

Health surveillance

Health surveillance for occupational disorders is often the job for an occupational health service, although some of the less technical procedures such as checking for signs of dermatitis, can be undertaken by people trained to recognise signs of the condition, e.g. supervisors. Health surveillance is not usually possible from primary care. However, a GP should have a basic understanding of the principles involved. The aim of health surveillance is to identify early signs or symptoms of disease, to inform the adequacy of measures taken to reduce risk of disease and to give employees information on their health status. Collection of information on the distribution and incidence of diseases (epidemiological data) is a secondary aim.

Health surveillance may be *statutory*, i.e. required by law. Examples include medical examinations and record keeping under the Ionising Radiations Regulations (1985), the Control of Asbestos at Work Regulations (1987), the Control of Substances Hazardous to Health Regulations (1999) and the Control of Lead at Work Regulations (1998). Particular medical investigations are required or advised under these Regulations and such surveillance can only be carried out by an Employment Medical Adviser or 'Appointed Doctor'. The Regulations allow the Health and Safety Executive (HSE) to appoint suitably experienced and qualified doctors to carry out health surveillance for employers.

Take workers with asbestos as an example of health surveillance:

Workers with asbestos (usually workers involved in asbestos stripping and removal) must have a two yearly medical examination which includes specific examination of the chest. The basic details of the employee's job, their exposure to asbestos, smoking habits and respiratory protection used are notified to the HSE. The purpose of these regular medicals is to identify signs of asbestos-related disease early and to advise the individual about whether to continue to work with asbestos.

Making an enquiry into smoking habits is an essential part of health surveillance for asbestos workers because smoking greatly increases the risk of lung

continued opposite

cancer associated with asbestos exposure, and the guidance actually states that part of the purpose of the health review is to advise the individual about smoking.

The signs and symptoms that you might expect in someone suffering from asbestosis are: dyspnoea on exertion with finger clubbing, which is slowly progressive (over many years). Bilateral, fine, high pitched crackles which do not clear on coughing may be present at the end of full inspiration and heard in both bases and lower axillae. Lung function tests would show a restrictive pattern with impaired gas transfer.

Health surveillance under the Control of Substances Hazardous to Health Regulations 1999 (COSHH) gives details of the levels of surveillance that may be necessary for workers with hazardous chemicals or biological agents. These range from the keeping of a simple record of exposure, through enquiry about symptoms, biological monitoring (blood, breath or urine samples) to a full medical examination by an Appointed Doctor or Employment Medical Adviser.

Health surveillance may also be required where there is a residual risk to the employees' health after all control measures have been taken. The Management of Health and Safety at Work Regulations 1999 and Approved Code of Practice (ACOP) lays down the following criteria for determining whether health surveillance is required:

1 there is an identifiable disease or adverse health condition related to the work concerned
2 valid techniques are available to detect indications of the disease or condition
3 there is a reasonable likelihood that the disease or condition may occur under the particular conditions of work
4 surveillance is likely to further the protection of the health of the employees to be covered.

This could be interpreted in practice as requiring health surveillance for exposure to many different hazards, including exposure to excessive noise and vibration, which commonly cause adverse health effects, but are not subject to their own statutory provision. The HSE has published guidance to explain when and how health surveillance might be carried out in such circumstances.[1]

Exposure to excessive noise is one of the criteria in the management of the Approved Code of Practice of the Health and Safety at Work Act for determining whether health surveillance is required. Undertaking health surveillance on a workforce exposed to noise (e.g. metal press-shop operators or foundry workers) entails pre-employment and regular audiometry. The technique can be carried out fairly easily, and can be done in the workplace as long as the audiometry equipment is properly maintained and calibrated and the test is carried out in quiet conditions (such as in a sound-proof booth).

The main diagnostic feature of noise-induced hearing loss (NIHL) on an audiogram is a dip in threshold of hearing at 4 kHz, with recovery of hearing threshold in the higher frequency. This '4 kHz dip' can be masked by the changes of presbyacusis; but it is worth looking for the dip if you are assessing an audiogram of someone who has worked in noisy conditions, and seeking specialist ear nose and throat advice if you suspect that there is an NIHL.

The environmental approach

Occupational hygiene

The handbook of the British Occupational Hygiene Society defines occupational hygiene as the applied science that is concerned with identification, measurement, appraisal of risk and control to acceptable standards of physical, chemical and biological factors arising in or from the workplace which may affect the health or well-being of those at work or in the community.

Investigating a suspected link between an occupational exposure and disease, or level of risk, may require the measurement of the hazard in the workplace against published exposure standards. Various measuring tools are used, including light and noise meters, and a range of sampling devices for gases, vapours, fumes and dusts.

▼

Occupational hygiene is a specialist area.

General environmental levels can be assessed by using a static sampler – a pump which draws air through a filter or absorption medium at a fixed rate. The amount of contaminant collected can then be measured (by weight in the case of dust or by gas chromatography for volatile chemicals) and airborne levels derived. A more simple method for gases and volatile chemicals is the use of indicator tubes. These glass tubes have an indicator medium, specific to the chemical under test, which changes colour as air is drawn through the tube with a hand-held pump. By drawing a fixed volume of air, levels of chemical in the air can be read directly from the colour change.

General environmental levels are important, but we are usually more concerned with exposures to individuals who move around and may be exposed to varying levels over a working shift. Hygienists have a variety of personal sampling devices, usually small pumps with collecting heads that can be attached near the worker's breathing zone.

Controlling risk

Risk assessment is not an end in itself, but should lead to the reduction or control of the risk. The principles of risk control are simple: first identify the hazards, then ask if they can be eliminated completely. Why handle a hazardous substance if you can eliminate it or substitute something less hazardous? Protective equipment should be a last resort, and the hierarchy of controls for all hazards is therefore:

1 *elimination* of the hazard
2 *substitution* for something less hazardous
3 *enclosure* of the process to reduce contact
4 *ventilation* – local exhaust ventilation may reduce exposure to volatile chemicals or fumes
5 *personal protective equipment*, e.g. respirators, goggles, ear muffs.

These principles apply to chemical, physical and biological hazards. The law requires that workers are educated about the hazards that they face and are trained to reduce risk.

The population approach to occupational health

Epidemiology is the study of the occurrence of disease in human populations. Epidemiological techniques can be applied to working populations, to try to answer the question of association between health effect and exposure to hazard. Much of this line of approach is not directly relevant to general practice, but you should be aware of the issues so that you understand why you might be approached for information or help.

A workforce is a fairly well-defined 'population' where exposures to specific hazards will vary from person to person. Other populations can be identified (e.g. a population of workers exposed to a specific hazard), depending on the problem being investigated.

The starting point for an epidemiological approach to an occupational health problem is the collection of data on health effects. These include data from death certificates, health surveillance records, sickness absence records, pension scheme records and primary care or hospital records. The national collection of data relating to occupational diseases is rather poor. There are several ways of finding accident and sickness data, including:

1 RIDDOR. The Reporting of Incidents, Diseases and Dangerous Occurrences Regulations 1995[2] require employers to report to the HSE any death or major injury at work, and to report certain dangerous incidents and occupational diseases. Injuries to be reported include fractures (excluding a bone in the hand or foot), amputations, loss of sight or penetrating injury to an eye, injury from electric shock, loss of consciousness from lack of oxygen and acute illness requiring removal to hospital or immediate medical attention. Any injury which results in the person taking more than three days off work or being unable to do their normal work for three days, must also be reported.

   The HSE estimates that only about a third of all eligible incidents/ diseases are reported which unfortunately, severely limits the usefulness of the data.

2 SWI Survey. The Self-reported Work-related Illness Survey 1995 was developed from the original 1990 'trailer' questionnaire on occupational ill-health in the Labour Force Survey. A representative sample of households in England and Wales was questioned on the incidence of occupational illness and accidents in the preceding year. The data from these surveys relies on self-report but when individual reports were checked with respondents' GPs, the latter provided a relatively high level of agreement.

3 OPRA, EPI-DERM and SWORD. The Surveillance of Work-related and Occupational Respiratory Disease (SWORD) reporting scheme was established to collect information on new cases of occupational lung disorder presenting to chest physicians and to occupational physicians. This scheme was followed by EPI-DERM, and more recently by the Occupational Physicians' Reporting Activity (OPRA), run from the University of Manchester's Centre for Occupational and Environmental Health. These schemes collect information on new cases of skin disease, musculoskeletal

disorders, hearing loss and other disorders presenting to a sample of dermatologists and occupational physicians, with the aim of improving information on the incidence of important occupationally associated problems.

4   Prescribed diseases. The Department of Social Security keeps data on the number of claims for 'prescribed diseases' under the Industrial Injuries Scheme. These give some indication of the incidence of certain conditions, and the cases are all validated by medical assessment. However, these data will be very dependent on the 'claims climate' and on rules determining entitlement to benefit. They underestimate the real incidence of new cases of disease and only give a guide to the absolute lower limit of the number of serious cases of industrial disease, rather than the true prevalence.

5   Death certificates. Death certificates recording death caused by asbestosis or mesothelioma are passed from the Office for National Statistics to the HSE for epidemiological study.

6   Periodic surveys. The ten yearly morbidity statistics from general practice and Decennial Supplements on Occupational Mortality contribute further information on occupational disorders.

7   Statutory health surveillance. Results of blood lead measurements under the Control of Lead at Work Regulations are recorded by the HSE.

The overall picture suggests that nearly 6% of adults reported suffering from a work-related illness in the 12 months to Spring 1990.[3] This implies a total of 2.2 million cases of work-related illness in England and Wales in a 12-month period. Of these cases 1.3 million were considered to be 'caused' by work (with a margin of error of 50 000 either way – 95% confidence limits). Not all these cases were serious; nearly half of the cases reported that they had not taken time off work, but the overall amount of lost working time was in the order of 13 million days.

Data on exposure to hazard are also likely to be incomplete or biased. It is rare to have a working population where full occupational hygiene and health surveillance records are available, and individual or group exposures are often badly documented.

Epidemiological investigations

An epidemiological investigation usually starts with a question – expressed as a 'null hypothesis', such as 'there is no association between occupational asthma and exposure to solder fumes'. Any study aimed at confirming or denying the 'null hypothesis' should be valid, feasible in terms of time and money and efficient.

There are several types of study:[4]

- *cross-sectional* in which a 'snap-shot' view is taken of the incidence or prevalence of disease in a population
- *historical cohort* in which the health status of a cohort of workers exposed in the past to solder fumes for example, would be compared with the health status of a comparable non-exposed cohort
- *prospective cohort* in which the health of a cohort of workers presently exposed to solder fumes for example, would be followed up for a certain length of time
- *case-referent* in which cases of asthma for example, are identified, matched with controls or referents, and the exposure history of cases and referents established to see if there are differences.

As the object of a study is to see if there is any association between the exposure and the health effect, certain questions are applied to the results:

- is the disease or disorder more common in a particular group (cases or referents), and if so, how large is the difference?
- has the association been described before in other relevant populations?
- does the exposure always precede the onset of disease?
- are there experimental (animal) data which confirm the likely pathogenesis?
- is there a dose–response relationship between exposure and disease?
- is the causality theory biologically plausible?

References

1   Health and Safety Executive (1999) *Health Surveillance of People at Work.* HSG 61. HSE Books, Sudbury.

2   Health and Safety Executive (1995) *A Guide to the Reporting of Injuries, Diseases and Dangerous Occurrences Regulations 1995.* L73. HSE Books, Sudbury.

3   Health and Safety Executive (1999) *The Costs to Britain of Workplace Accidents and Work Related Ill-health* (2e). HSE Books, Sudbury.

4   Rose G, Barker DJP and Coggon D (eds) (1995) *Epidemiology for the Uninitiated* (3e). BMJ Publishing Group, London.

Reflection exercises

If you want to consolidate your learning, you need to spend some time thinking about how what you have read applies to your own practice. Why not do one, or both, of the following exercises?

Exercise 1: Work through these two case histories describing how you would approach these two conditions using what you have learnt from reading the material in this chapter. You might read the extracts from the Occupational Health Advisory Committee's report and recommendations at this point to broaden your perspective of occupational health support (*see* Appendix 2).

1 Dawn, a factory worker

A 42-year-old lady comes to see you, complaining of pain in both wrists. No other joints are affected, and the pain seems to be largely around the extensor tendons of the thumb and fingers. There is no crepitus but she is tender over the tendon sheaths. Her notes record that she works at a very large local chocolate factory.

(a)   *What else do you need to know about her job?*

She describes the tasks that she performs now, which involve hand-packing chocolates into boxes.

(b)   *What are the treatment options?*

(c)   *Who would you try to contact about this problem, and what would you ask them?*

(d)   *What options for managing this problem do you think the occupational health department will consider?*

This part of Exercise 1 might take 1 hour.

2 Greg, a shipbuilding worker

A 58-year-old man attends your surgery with his wife. She tells you that she is fed up with her husband turning the TV volume up, but he won't accept that he is deaf. He grudgingly admits that he has real problems hearing people speak, especially in noisy surroundings. There is no obvious problem with his ears, and no family history of deafness. He appears to have a bilateral sensorineural hearing loss.

(a)   *What do you want to know about his occupation(s)?*

He has worked in shipbuilding all his life, from leaving school at 15 years old. You refer him for an ENT opinion, but in the mean time, he asks you what he could do about getting 'compensation' if his problem turns out to be caused by his work.

(b)   *What compensation and other benefits are open to him, and how would you suggest he goes about claiming?*

Six months later, he comes to thank you for sorting out his hearing, and says in passing that he has noticed that he is getting short of breath. He doesn't look well, and when you examine his chest, he has some dullness at the right lung base.

(c)   *What further history do you take?*

He coughed up blood on one occasion a couple of weeks ago, but dismissed it as a 'strain'. His chest X-ray shows an opacity and effusion at the right base, with bilateral diffuse pleural thickening and you refer him for an urgent medical opinion, and give him a Med 3 for 4 weeks.

(d)   *What diagnosis would you put on the Med 3?*

When you see him for review, he has been told the diagnosis by the chest physician, and treatment has been discussed. You give him a further Med 3 for 8 weeks. Two weeks later, the occupational physician at his place of work writes to you with the patient's written consent, asking for further details of his diagnosis and prognosis.

(e)   *What options might the occupational physician be considering for the patient's future employment?*

(f)   *What are the likely pension, compensation and social security implications of this situation?*

This part of Exercise 1 might take 1 hour.

Exercise 2: Keep a log and make a record every time a patient consults you where the occupational history is a factor in the reason for consultation. Keep the log for a week.

(i)   How many patients consulted you with symptoms linked to their occupation in the week of observation? What proportion of patients consulting you was this?

(ii)   Did any patterns emerge; e.g. of people consulting from one particular workplace, or of common occupationally related conditions?

(iii)   How many were able to remain working?

(iv)   In how many of those who were able to continue at work, (a) were you able to give treatment or advice that should enable them to eliminate or prevent the hazard at work, (b) minimise the hazard – or (c) there was no treatment or advice you could give them from primary care that would make any difference to the hazard at work or its effects.

This Exercise should take about 2 hours if you work full-time; or proportionately less if you work part-time.

CHAPTER FOUR

Fitness for work

Start with the occupational history when determining a person's fitness to do particular jobs. Find out fairly precisely what tasks are performed, what hours and shift patterns are worked, what the working/environmental conditions will be like.

The sorts of questions and issues you should be considering are:

- Is the person medically (physically and mentally) fit for the job and is the job likely to be particularly harmful to the individual?
- What are the physical requirements of the job-related tasks, for example in the form of a 'Physical Demands Analysis' (PDA)?
- What is the individual's functional capacity against a specific job requirement or PDA? This can be used as the basis for matching the person to a specific job or a general type of work and may include for example, suggestions for aids and adjustments and may encompass not only capacity but skills and motivation.

Sophisticated employers produce a written job description with details of the physical and psychological demands for each task. This can help the doctor detect any mismatch between the patient's health and the job demands. Unfortunately, this approach is rare, and you may have to go on what the patient tells you or contact the employer for further information if you are asked difficult questions – all very time-consuming!

Special problems of fitness for work are:

Food handling

There will be a number of 'food handlers' in your practice population and you may be asked for advice on possible infection issues. There is no definition of a 'food handler' in UK or EU legislation and the current guidance is aimed at three categories of persons:

- those employed directly in the production and preparation of foodstuffs including the manufacturing, catering and retail industries

- those undertaking maintenance work or repairing equipment in food handling areas
- visitors to food handling areas, including enforcement officers.

Workers who handle only pre-wrapped, canned or bottled food, or those involved only in agriculture or harvesting are not considered to be food handlers.

Many employers and employees still believe that some sort of pre-employment 'medical' including stool testing is necessary to ensure fitness to work as a food handler, but current guidance is aimed at educating and training in food hygiene rather than medical testing. However there are some medical contra-indications to working as a food handler.

▼

Take care with food handlers.

Medical conditions that might prevent someone working as a food handler

- a history of enteric fever (infection with *Salmonella typhi* or *Salmonella paratyphi*)
- a recent history of diarrhoea and/or vomiting
- chronic skin disease affecting hands, arms or face

continued opposite

- boils, styes or septic fingers
- discharge from eye, ear, or gums/mouth
- recurring bowel disorders
- recent contact, at home or abroad, with a possible typhoid/paratyphoid carrier.

Excluding these conditions in a prospective food handler gives an opportunity to reinforce education and training on basic personal hygiene. Such health promotion is much more useful than issuing a 'freedom from infection' certificate, which may still be requested under certain circumstances.

It is the employer's responsibility to enquire from the prospective employee about possible infection risks, and there is no evidence that medical certification prevents the spread of infection from an infected food handler as it provides only 'snap-shot' information about the prospective employee's health status. Nevertheless, if an employer believes that there is a relevant history of infection or an impediment to handling food, he may ask the GP to provide a medical certificate after appropriate enquiries and investigations. There is no provision for this to be done under the NHS, so the GP may charge a fee.

There are some UK food regulations which require medical certification of fitness for work, including the Dairy Produce (Hygiene) Regulations 1995, the Fresh Meat (Hygiene and Inspection) Regulations 1995 and the Egg Products Regulations 1993.

A food handler who has an acute diarrhoeal illness can return to work subject to the following requirements:

- no vomiting for 48 hours once any treatment has ceased
- bowel habit has returned to normal for 48 hours either spontaneously or following the cessation of treatment with antidiarrhoeal drugs
- good hygiene practice, particularly hand-washing, is observed in all circumstances.

The investigation and management of suspected cases of enteric fever is usually carried out by the local authority, but the GP may be involved in treatment. Essentially, the employee who has had *Salmonella typhi* or *paratyphi* infection is fit to work only after six negative stool specimens at intervals of not less than 14 days. If any specimen is positive, further treatment (usually with a quinolone antibiotic) may be required and the sampling programme will begin again.

In the case of verocytotoxin-producing *Escherichia coli* (VTEC) such as *E. coli* O157:H7 being implicated in diarrhoea in a food handler, they should be excluded from work until their bowel habit has been normal for 48 hours and two consecutive stool specimens taken 48 hours apart have been negative.

If a food handler has diarrhoea associated with *Campylobacter*, they can be treated with appropriate antibiotics and allowed to return to work once their bowel habit has returned to normal, and as long as their personal hygiene is of the highest standards. Negative stool specimens following treatment are not essential – although they may be reassuring – as they do not prove that the patient is completely free of the organism.

Employees with hepatitis A should remain off work until 7 days after the onset of symptoms, usually jaundice. Contacts of a hepatitis A case need not be excluded from work as long as they practise good personal hygiene. Vaccination of food handlers is not considered to be necessary.

Chronic gastrointestinal illnesses such as Crohn's disease are not a contraindication to food handling, nor is the presence of a colostomy or ileostomy, so long as good personal hygiene is practised. Change in bowel habit must be considered to be infectious and the employee excluded until symptoms have resolved. Chest disease, including tuberculosis, is not a contraindication to work with food, except on grounds of general fitness.

Further information on the medical aspects of food handling is contained in the publication: *Food Handlers – Fitness for Work*[1] prepared by an expert working group convened by the Department of Health.

Professional driving

Britain has one of the lowest fatality rates for road traffic accidents (RTAs) per 100 million vehicle kilometres travelled apart from the USA. Nevertheless, legislation requires a higher standard of medical fitness for Group II (larger vehicle) drivers, which is also recommended for other professional drivers such as ambulance, police and taxi drivers. Drivers of large goods vehicles (LGVs) have about three times the RTA fatality rate per mile travelled compared with car drivers.

Responsibility for determining the fitness to drive of an individual rests with the Driver and Vehicle Licensing Authority (DVLA).[2] Licence holders have a responsibility to inform the DVLA of any medical condition which might affect their driving, and a doctor may be asked to provide a medical report, but will not be asked for an opinion on fitness to drive. The DVLA certifies individuals as fit to drive Group II vehicles and a doctor only undertakes the relevant medical examination and completes the form D4. Doctors may be asked to give opinions on fitness to drive taxis (for local authority licensing) or for insurance companies. Due care and skill must be exercised where opinions on fitness are given and there is a potential legal liability. Up-to-date guidance can be found in the DVLA's booklet for medical practitioners.[2]

Driving a mobile crane after a coronary artery by-pass graft

Drivers of mobile cranes now have to hold at least a C1 driving licence (if the crane is between 3.5 and 7.5 tonnes) or a Group 2 (Category C) licence for cranes exceeding 7.5 tonnes. The medical criteria for categories C1 and C differ slightly, and advice should be sought from the DVLA. Essentially, driving should cease for at least 6 weeks after the surgery, but as long as the patient is then free from angina and can complete a standard exercise test, driving may be resumed.

Some of the main medical problems for Group II drivers include:

*   *Cardiovascular disease*
    Angina: recommended permanent refusal or revocation of licence
    Myocardial infarction: driving to cease for 3 months; return to driving allowed when symptom-free, and able to complete a standard exercise test
    Arrhythmia: refusal/revocation if the arrhythmia has caused or is likely to cause incapacity (including systemic embolism)
    Pacemaker: disqualified from driving for 3 months; relicensing may be permitted thereafter unless there are other disqualifying conditions
    Hypertension: revocation/refusal for casual BP $> 200/110$, or established BP $> 180/100$, until satisfactorily treated
*   *Diabetes mellitus*
    Insulin dependent (type 1): permanent refusal/revocation
    Diet and tablets: review of medical conditions, with revocation if they develop relevant disabilities (e.g. eye problems)
*   *Epilepsy*
    Allowed to drive Group II if free from fits for 10 years (on no medication) and free from 'a liability to epileptic seizures', e.g. from a structural intracranial lesion
*   *Vision problems*
    New applicants must have a visual acuity of at least 6/9 in the better eye and at least 6/12 in the other eye (corrected). They must also have an uncorrected visual acuity of at least 3/60 in each eye tested separately.
    Present drivers are allowed to drive even if effectively monocular, or if their visual acuity only reaches a pre-1983 standard. But they must be able to certify that they have been driving regularly (at least once a month, for 6 months in the past 5 years) and are accident-free in the past 10 years where eye-sight might have been a factor.

Fitness for overseas travel: an example

A 32-year-old married engineer consults the practice nurse for advice. He is being seconded for 12 months to an oil exploration project in East Africa. What factors does the nurse need to think about when advising him on the medical aspects of the trip?

Consider whether his employer has an occupational health service to help him prepare or if there is an occupational health service where he is going.

If the patient will not have access to good quality emergency medical facilities he might take a pack of sterile needles and intravenous giving sets, etc. He could register with an organisation like the Blood Care Foundation, which makes fully screened blood available to its members anywhere in the world in an emergency. Road traffic accidents are one of the highest causes of morbidity for workers in this sort of situation. Ask him to think about possible evacuation procedures in the event of sudden illness or injury.

The patient may have medical contraindications to travel – chronic disease that is liable to sudden exacerbations, or chronic anxiety or depression, etc. Discuss if he is on any long-term medication, and if so, how will he maintain supplies. He will need his vaccinations and malaria prophylaxis sorting out – his employer should be prepared to pay for appropriate prophylaxis. He might take an emergency supply of antibiotics for acute diarrhoeal illness, such as ciprofloxacin.

Enquire how his family feel about him going away for 12 months and whether his wife will join him at any stage, in which case she will need her own medical preparation.

He should be counselled about the psychological stresses of doing a physically demanding job in a different culture, away from family support for a year. He should also be warned about sexually transmitted diseases – particularly HIV.

References

1   Department of Health (1995) *Food Handlers – Fitness for Work. A report of an expert advisory panel.* HMSO, London.

2   Drivers Medical Unit (1999) *At a Glance Guide to the Current Medical Standards of Fitness to Drive; for medical practitioners.* Driver and Vehicle Licensing Agency, Swansea.

Reflection exercises

If you want to consolidate your learning, you need to spend some time thinking about how what you have read applies to your own practice. Why not do one, or all, of the following exercises?

Exercise 1. Composing guidelines for a consistent approach to fitness for work in general practice.

Draw up an information sheet for your practice describing best practice in considering whether patients (of working age) are fit to work. Incorporate the issues and concerns around mobility, medication, safety, lifting, driving and freedom from infection – in each case if it is relevant to an individual patient's job or condition. Do this exercise around four conditions:

- a patient who has had a myocardial infarction
- a patient with food poisoning
- a patient who has had an abdominal operation
- a patient who is depressed.

The outcome of this exercise should be a more consistent approach to fitness for work – by the same practitioner and between different GPs. *This Exercise might take 2 hours.*

Exercise 2. Fitness for work – a firefighter.

Think through the likely job demands, hazards and risks of a firefighter, and try listing some of the definite, probable and possible medical contraindications to firefighting. Use some of the principles you have learnt from the material in Chapter Three. *This Exercise might take 1 hour.*

Exercise 3. The Occupational Health Advisory Committee report states that 'Primary care is an essential provider of occupational health support' (*see* Appendix 2). Describe the *strengths, weaknesses, opportunities* and *threats* for you as a practice in providing occupational health support to your individual patients, and any small or medium enterprises or businesses in your practice patch. Draft an action plan to strengthen your capability in providing occupational support, from the practice. *This Exercise might take 2 hours.*

CHAPTER FIVE

Ethics, communication and confidentiality in occupational health practice

Medicine has many ethical codes and rules, from the Hippocratic Oath to its modern restatement, the Declaration of Geneva (1947), but these codes stress the principle of *primum non nocere* (above all, do no harm). They also stress the doctor's responsibility for the individual whilst essentially ignoring the interests of populations or society at large.

In the 21st century we have now moved beyond the scope of the Hippocratic Oath. Medical practice has expanded dramatically and improving the health of the individual (or even the population) is no longer the single aim. Medicine is used to 'improve' the quality of life or intervene in situations that do not directly relate to the health of a patient, for example in contraception, sterilisation, *in vitro* fertilisation and cosmetic surgery. There are problems as to what constitutes health and which attributes of health are obtainable and at what cost.

Guidance from the Faculty of Occupational Medicine[1] points out that doctors have three forms of contact with patients: in the traditional doctor–patient relationship, as an impartial medical examiner reporting to a third party, and as a research worker enquiring into the causes of disease. Occupational physicians fulfil all these roles, and restoring or maintaining health may only be a small part of the occupational physician's job. A GP or nurse may spend most of their practice in a therapeutic relationship with their patients. The potential for conflicts of interest is therefore greater for an occupational physician.

Conflicts of interest

Occupational health services must be seen to be impartial. To command the respect of managers, employees and trade unions, an occupational physician has to be seen to be a professionally independent adviser, concerned with the health of the employees, but not always an advocate for the patients – as their GP might be.

If the employee believes that the doctor is giving friendly and confidential advice in a traditional doctor–patient relationship, and subsequently finds that the consultation is used to advise management about fitness to work or potential future problems, confidence in the independence and impartiality of the occupational health department can be destroyed. Occupational physicians must also ensure that others (managers, employees, trade unions, and professional colleagues) understand the nature of their role in every situation. The only way of ensuring that this impartiality is established and reinforced is to behave in an open and honest manner in all dealings with management and employees.

▼
The employee's interests should be paramount.

Confidentiality

Occupational physicians are under the same constraints of medical confidentiality as other doctors. But maintaining the confidentiality of occupational health records is not quite as straightforward as that of hospital or general practice notes. There are several key areas that cause concern:

- ownership and guardianship of records
- security of records
- access to records
- disclosure of records.

Ownership and guardianship of occupational health records

The owner of the paper, the folder and the filing cabinet is likely to be the employer for whom the occupational physician works. This can lead to problems if the employer decides he wishes to see the records; there have been situations where the doctor is ordered to hand over records to personnel departments or other managers. This would be a clear breach of confidentiality and must be refused. The content of the record is the property of the author and must not be handed over without the informed, written consent of the individual to whom the record relates. Occupational health records should therefore be under the care of an occupational health professional (doctor or nurse) wherever possible.

Security of records

All occupational health records should be secure and not open to unauthorised interference or inspection, just as for general practice records. Records maintained on computer or in a manual retrieval system are now subject to the provisions of the Data Protection Act 1998. It is particularly important for health professionals who wish to store and use 'sensitive personal data' for research purposes to understand the provisions of the Act in relation to subject consent and subject access. Further details can be obtained from the Data Protection Commissioner (tel: 01625 545745. email: mail@dataprotection. gov.uk). This applies equally to computerised records which should be protected from viruses and unauthorised network or telecommunication access.

Access to records

Under the Access to Health Records Act 1990 an individual has had the right of access to records made after November 1991 'in connection with the care of that individual' with an occasional exception to this rule. Occupational health records may come within that definition.

Disclosure of records

Occupational health physicians are often asked to release records to lawyers in cases of litigation or to industrial tribunals in cases of dismissal on the grounds of ill-health, poor performance or excessive absenteeism. The rule is the same in every situation: the individual's fully informed and written consent must be obtained before release. Occupational physicians usually discuss the extent of the 'consent' with the individual, to ensure that they are truly 'fully informed' and that their wishes are respected.

Rarely, legal advisers may ask for disclosure of very specific information only. For example, they may ask for a recent audiogram that is consistent with a diagnosis of noise-induced hearing loss, but not for earlier ones showing perhaps a similar pattern. In this situation, the occupational physician should consider advising the solicitors of both parties that all relevant records should be available to both sides, and that the individual's consent (or a court order) should be obtained.

Fitness for work and health screening

Pre-employment assessment

When a doctor is asked to assess fitness for a particular job at the pre-employment stage either by medical examination or health questionnaire, the primary responsibility is to the employer. Any information given to the employer is only in terms of 'fitness to work'.

Occasionally an occupational health service will ask the individual's GP for a report at the pre-employment stage, justified by anxiety over potentially concealed but significant illness. The Allitt case[2] is typical, in which the deaths of children in the care of a psychiatrically ill nurse might have been averted if the occupational health department had had access to her previous history. Whenever asking for a report in these circumstances, the occupational physician should be clear about why the information is necessary and should seek informed written consent, rather than rely on a proforma consent which a job applicant is hardly likely to refuse to sign.

GPs are likely to get such requests for information from time to time, and their reports will be used to help the occupational health service to advise management, bearing in mind the provisions of the Disability Discrimination Act. Any report therefore needs to be based on a real understanding of the job concerned. The onus is on the potential employer who is requesting the report to make the GP aware of any specific requirements of the job which may have a bearing on their patient's fitness.

Management referral after absence attributed to sickness

The occupational physician's responsibility is primarily to the employee when assessing someone with a poor sickness absence record. Care must be taken to ensure that the employee understands the doctor's role, and that the doctor is not being used by management to put pressure on an employee. The control of absence attributed to sickness is a legitimate interest of occupational health departments, but is a *management* responsibility. In the case of an employee referred under specific disciplinary procedures (e.g. drug and alcohol abuse policies), the doctor must make it clear to the individual precisely what the purpose of the referral is and what management expects.

Ill-health retirement

Occupational physicians are alert to the possibility of being used to shed un-wanted labour via the ill-health retirement route rather than via redundancy. They should understand the provisions of the company's pension scheme and should be familiar with the state benefits which may apply. Occupational health doctors should always give impartial advice and not be swayed either by an employer's pressure or the employee's wishes. They should also be pre-pared to collect supporting evidence of being permanently unfit to work (with consent) from GPs and hospital specialists. This is one area where com-munication with the GP is vital in order to ensure cooperation and support, and it is essential that the occupational physician and GP remember that a person who is unfit for a particular job is not necessarily unfit for all work – as discussed before. It is important that patients are not wrongly advised to think that if the GP is prepared to issue sickness absence certificates they will continue to get a state incapacity benefit.

Apart from the GP being asked whether to support ill-health retirement for one of their patients (and this usually implies continuing to issue sickness absence certificates) he or she may very occasionally be asked in confidence about the prognosis of life-threatening illness. Many pension schemes have a clause allowing members to waive their pension in favour of a large lump sum on retirement. This option is not usually worthwhile but if someone knows that they have only a few months to live, signing a 'commutation of pension' may give their family a very substantial lump sum. Clearly this may be a delicate matter because the patient may not know (or may not want to know) their prognosis. If you feel that they should not be told that they have a very poor prognosis even if this means a loss of pension, this information is very important to the occupational physician and the pension fund trustees, who will want to give the patient the best possible financial outcome.

Research

Occupational health services may be asked to contribute to research projects and should be careful to ensure that any project has Local Research Ethics Committee approval. They should check that the procedures for informed consent are clear and that any refusal of an individual to take part does not interfere with their relationship with the occupational health department. A patient's GP may be contacted before the research to ensure that there are no medical contraindications unknown to the research team. Research results should be explained to individuals and (if appropriate) to the whole work-force, ensuring that no individual can be identified from the results.

Communication and relationships with others

Communication of information to the individual

An employee has certain rights under the Access to Health Records Act 1990 and the Access to Medical Reports Acts 1988, and the principles embodied in these two Acts are regarded as good occupational health practice even when they may not formally apply. The doctor should be careful to explain everything to the individual, ensuring that the doctor's role is made clear and that the employee is aware of what is to be reported to management and others. As previously mentioned, the Access to Medical Reports Act 1988 does not apply to reports requested by the Department of Social Security or the Benefits Agency.

Communication with management

Advice should always be given in terms of limitation of function and fitness to perform specific tasks and clinical details should not be given. If it is felt that further specific details would be helpful (e.g. an insulin-dependent diabetic may want his work-mates and first aiders to understand the condition) then written consent should be obtained before divulging the information.

In the rare event of a condition being discovered which makes the individual unfit to do certain tasks involving the safety of other workers or the public, the doctor must explain the reason why disclosure is vital. If the individual refuses consent, the doctor should then seek defence association advice on breaching confidentiality on the grounds of over-riding public interest.

Communication with professional colleagues

Good communication between GPs and occupational health departments is extremely valuable but sadly not always achieved. Good communication should include not only formal requests for information (with consent, and subject to the Access to Medical Reports Act) but also informal discussions on the possible hazards to health in the workplace and the role of the occupational physician – still very often misunderstood!

From the other side of the fence, the occupational physician will normally keep the employee's GP informed of specific problems and of health events affecting the patient, but only with the individual's consent. Referrals to hospital consultants will normally be made by the GP except in an emergency, and the occupational physician should be careful to avoid being used as a 'second opinion' on general health matters by employees.

Occupational physicians work with nursing staff, who are themselves bound by codes of ethical conduct. It is important that the medical and nursing staff understand each other's roles, particularly where the doctor is part-time and the nurse has managerial responsibilities for the occupational health department. The GP, the reception staff and the practice nursing staff may therefore have to communicate directly with an occupational health nurse where there is no full-time physician; you should expect that the nurse will abide by all the same rules of confidentiality as a physician.

Observance of both general ethical codes and the specific advice available via the Faculty of Occupational Medicine, the BMA and the defence organisations is vital to avoid ethical and legal disasters, and to maintain the credibility of the specialty of occupational medicine.

References

1   Faculty of Occupational Medicine (1999) *Guidance on Ethics for Occupational Physicians* (3e). Faculty of Occupational Medicine, London.

2   Department of Health (1994) *The Allitt Inquiry: report of the independent inquiry relating to the deaths and injuries on the children's ward at Grantham and Kesteven General Hospital during the period February to April 1991.* HMSO, London.

Reflection exercise

If you want to consolidate your learning, you need to spend some time thinking about how what you have read applies to your own practice. Why not do the following exercise?

Exercise 1. How would you respond to this letter from an occupational physician? Write up to 500 words pulling out the ethical issues that this case throws up – discussing the points raised in this letter and those that are not mentioned. Consider the ethical, communication and confidentiality issues from the three perspectives of the patient, the GP and the occupational physician.

Dear Doctor,

Peter Jones (12.3.69)

I attach a copy of this gentleman's written consent to contact you for a report. You will note that he wishes to see your report before you send it, under the Access to Medical Reports Act.

He is an electrical fitter, and I saw him today at the request of his supervisor, who has become increasingly concerned about his sickness absences and his performance. He has had 27 days off in the past nine months, mostly in one to three-day spells, which you may not be aware of. His supervisor has noticed that he smells of alcohol and he has made two quite serious mistakes at work, one of which destroyed a bank of electrical equipment.

I questioned him about his health in general, and about his alcohol intake in particular, and he was evasive, saying that he only drank at weekends. He then said that you are treating him for depression because of family problems.

Could you therefore tell me:

    What are his present medical problems? Are there underlying family or social problems?

What treatment is he presently receiving from you, and is he complying with your treatment?

Does he (in your opinion) have an alcohol problem?

Has he been referred to a specialist for further advice, and if so, what was the outcome?

Are there any other medical factors that are relevant at present?

My aim is to help him remain at work at present, and to overcome his present difficulties. Any information you give me will remain confidential to this department.

Yours sincerely,

This Exercise might take 1 hour.

CHAPTER SIX

Dealing with common occupational health issues in general practice

This chapter outlines a series of occupational health problems which will be commonly encountered in general practice to:

*   enhance the recognition of such problems and to improve effective management of them
*   enable practitioners to help people return to work safely and effectively.

There are ten sketches or scenarios. The scenarios are divided into two groups according to the two main facets of occupational health practice:

*   the effects of work on health and
*   ill-health effects on capacity to work.

Historically, occupational health (OH) has developed largely outwith the NHS structure, and is thus one of the least familiar specialties. The well-being of the individual patient is the dominant principle, enshrined in written ethical guidelines for both doctors and nurses in OH. Advising on the ordering of health and safety arrangements in the organisation employing the OH personnel is the secondary function. Although uncommon, there is potential conflict between the first and second duty. When it occurs, other practitioners may be drawn in. Thus the experience of GPs not otherwise engaged in OH practice may well be disproportionately that of involvement in advocacy or claimancy or both.

The scenarios which follow are set out in a specific order. Information gathered from dealing with one, should help you to address those that follow.

By the end of the chapter, you should have incorporated the occupational dimension into your way of thinking about patients' problems and their sequelae. The differentiation between the effects of work and the effects of health are not mutually exclusive, especially in more complex scenarios. Nevertheless, making the distinction is useful in deciding what you are dealing with and how you are going to handle the situation.

The effect of work on health

About 8% of transactions between doctors and patients in general practice have a major occupational component. The issues, ideas and principles established can be set down in a relatively simple set of headings.

- What work do you do? (Ramazzini's question – *see* opposite).
- What work have you done?
- What does/did it actually involve?

Work is a key defining characteristic of adult status and threats to that status should not be lightly made. Accurate diagnosis, consultation with the patient and informed advocacy are essential.

Scenario 1: occupational asthma

Chris S and his family have been with the practice for decades. His thick file is evidence of an eventful childhood and an anxious mother. He is now 28 years old and married with two children.

   Earlier in the year he came to see you with an intermittent history of cough and chest tightness over 3–4 months. He said it usually occurred at home, often in the evening and sometimes stopped him from getting to sleep for a bit. He said it didn't bother him all that much and was no worse than what he remembered having had as a child. He had been provided with a bronchodilator spray to be used as necessary on the assumption that he had experienced a recrudescence of childhood asthma. He had reported effective relief from this.

   Now he has asked to see you again following an inhalation incident at work when there was a chemical spillage. He had been admitted overnight to hospital because of this and you have the discharge letter in front of you. It reports that on admission he had the following spirometry results:

$$FEV_1 - 2.1 \ l \ (3.7) \qquad FEV_1/FVC = 50\%$$
$$PEFR - 250 \ l/min \ (480)$$

Figures in parentheses are predicted values. FEV_1 is forced expiratory volume in 1 second, FVC is forced vital capacity and PEFR is peak expiratory flow rate.

Overnight Chris had recovered to 3.21 FEV_1 and 410 l/min PEFR and, being symptom free, was discharged.

Chris reports that since he had the spillage exposure at work, his asthma has got more frequent and more severe although it has still responded perfectly well to the prescribed bronchodilator. He is also worried that the inhalation made him a lot worse than any of his three colleagues who were also involved. None of the others were admitted to hospital.

When you examine him, his lungs sound clear and he himself says he has no symptoms this morning. Lung function parameters are within 10% of predicted values. On questioning he identifies the spillage as an ammoniacal compound, pungent in odour, which is kept as a concentrate ready to dilute and added to cutting fluids in the machine-shop at work.

Questions you might ask Chris in order to help you with your diagnosis

Two sets of questions might clarify what is going on. One set relates to Chris' asthma, the other set to the spillage incident. The first and simplest question is that first posed by Ramazzini who wrote the first textbook of occupational medicine in 1700 and is thus known as the 'father' of the specialty. He proposed that doctors should always ask an individual what his or her work was. It is still a question not asked often enough!

Chris turned out to be a spray-painter who worked with isocyanates (e.g. '2-pot' paints), the commonest cause of occupational asthma. Other common occupations associated with asthma are bakers or flour workers, and experimenters or technicians working with laboratory animals.

Other useful questions about the asthma will be to check whether it occurs at weekends or holidays. Hobbies, especially those involving soldering (colophony) may also cause asthma. If the patient works shifts, the pattern of asthma may 'follow' the shift pattern. Finally, the resultant generalised airway reactivity can be checked by asking about the effect of challenges like cold air and exercise.

In relation to the spillage, questioning should elucidate the speed of onset of respiratory distress and its severity. Sequelae to be scrutinised will include severity, frequency and nature of symptoms before and after the spillage incident.

The differential diagnosis is:

*   asthma
        intrinsic
        occupational
        hobby

- airways lability
    generalised
    reactive airways dysfunction syndrome (RADS).

Features of Chris' illness which might lead you to conclude that he had occupational asthma

Typically occupational asthma is manifested *after* the work inducing it, at least at first. So, evening chest tightness with recovery by the next morning is a common pattern. As the problem gets more severe the asthma may encroach on the working day. This 'delayed' form may also be associated with an acute component presenting 'immediately' in the workplace, again often as the condition worsens. Alleviation/absence of symptoms at weekends and especially on holiday is also a very helpful diagnostic criterion. More complex patterns involving worsening over the weekends and delayed recovery are also seen and, unfortunately, a substantial minority of occupational asthma presentations are in this, or other ways, atypical.

Objective ways in which you might test your suspicions

Whilst the history of asthma linked to the occupational history is the main-stay of diagnosis, the story can often be usefully confirmed by ambulatory spirometry in a similar way to that used to assess other asthmatics. Classic-ally, peak-flow spirometry is done over a 4-week period taking readings two hourly during waking hours. The diagnosis of occupational asthma is made on a pattern of at least two work-related worsenings of PEFR and two recoveries out of exposure (15–20% + variation). Other hospital-based techniques include challenge to assess airways reactivity; skin prick tests; radioallergo-sorbent test (RAST); enzyme-linked immunosorbent assay (ELISA) for specific causal agents.

As a result of your investigations, you consider it highly likely that Chris has had occupational asthma for at least some months. Additionally he may have RADS.

Advice about work, prognosis, and treatment

The uniqueness of the occupational variant of asthma lies in its potential for total cure by avoidance of exposure. However, many studies have now shown that airways hyper-reactivity can and often does persist for years after removal from exposure. Therefore *early* detection and *complete* removal is the best option for the disease. Unfortunately it is often not the best option nor one that is acceptable to the individual. Thus, Chris as a skilled spray-painter, will not take kindly to the suggestion that he should discard this skill and the thicker wage-packet associated with it.

It may be that the first principle of occupational health control, avoidance or substitution (of a noxious agent), will have to give way to control to a low level (compliance to a so-called hygiene standard) or to the use of masks, etc., as personal protective equipment (PPE). Regular prophylactic or symptomatic treatment may also be required. COSHH (Control of Substances Hazardous to Health Regulations) require employers to carry out these control activities in a systematic way assessing the risks and monitoring exposures and health.

Unfortunately, agents causing occupational asthma do not fit too well into this structure even though, as in the case of isocyanates, a hygiene standard may exist and be enforced. Medical surveillance thus becomes a mainstay of management with the unfortunate effect that sensitised individuals may lose their jobs. Typically, in the spraying industry these people will wish to continue in the same employment despite the threat to their health and tend to drift into employment with smaller, less fussy and less well-regulated employers and thus be at greater risk. An unpleasant paradox.

Susceptibility to occupational asthma

A fairly constant finding in studies of occupational asthma are positive associations between asthma and a couple of possible predisposing factors such as atopy and smoking. Some industries (e.g. platinum refining) preclude atopics from employment. The general scientific consensus at present is that in general neither atopy nor smoking is a sufficiently sensitive or specific predictor for exclusionary purposes.

Further or prior reading

Health and Safety Executive (1994) *Preventing Asthma at Work*. HSE Books, Sudbury.

Hendrick DJ (1999) Work and chronic airflow limitation. In Baxter PJ *et al.* (eds) *Hunter's Diseases of Occupations* (9e). Arnold, London.

Newman Taylor AJ (1999) Occupational asthma. In Baxter PJ *et al.* (eds) *Hunter's Diseases of Occupations* (9e). Arnold, London.

Scenario 2: vibration white finger and hand–arm vibration syndrome (VWF and HAVS)

You have seldom if ever seen Kevin, although his wife Maureen and their innumerable children are frequent visitors to the surgery. He turns out to be a rather shy and reticent 39 year old.

Kevin tells you that Maureen has told him to come along because she doesn't like his cold hands when they are making love. Kevin is also anxious about his hands because they sometimes go numb and lately they have woken him up at night with pain and a disturbing tingling. He explains that he has always worked physically hard throughout his life and has never had difficulty getting off to sleep or staying asleep before. He confesses that he is anxious about the 'funny feelings' in his fingers because this is how his brother Damian started with multiple sclerosis.

When you question Kevin, he gives a history of 'some years' of short episodes of hand and finger blanching and coldness. The episodes last a few minutes. In the last year or two they have been accompanied by numbness and tingling and latterly this had been present even between the blanching episodes. He found he could deal with the episodes by warming his hands: 'it gives you a funny throbbing feeling when the blood comes back, doc'. To avoid attacks he now wears gloves or mittens in the winter and has given up walking the dog on cold days.

Initial differential diagnosis

The differential diagnosis is:

- Raynaud's disease (feet, ears, nose)
- previous hand trauma
- autoimmune conditions (especially scleroderma)
- rheumatic diseases
- atherosclerosis (especially Buerger's syndrome)
- hand–arm vibration syndrome (HAVS).

When you examine him in the surgery you find a fit man, well muscled with calloused and 'well-used' hands. The only signs that you can elicit are in the fingers. Although the hands and fingers are warm on initial examination, three minutes immersion in cold water produced a blanching attack 'just like those I get normally, doc'. (This is only an indicator – not a very reliable test.)

Light touch and pinprick perception are diminished in the fingertips at initial examination. As an incidental observation you notice that Kevin sometimes has difficulty hearing your questions.

Questions that might help to refine the differential diagnosis

Hopefully you will have thought straight away of asking about Kevin's work. Eliciting an occupational history in Kevin's case will tell you all you want to know although bringing out that history may take quite a lot of skill on your part. Kevin will not necessarily know what aspect of his work is important to you to make a diagnosis. He will thus tell you that he 'used to work on the roads' and now works 'fettling'. Because he knows what he means, he may assume that you do too. Similarly he may well be unused to thinking generically about classes of work or tools. Thus he may not respond to questions asking him about work with 'vibrating tools'. Useful questions here include:

- what equipment did you/do you mainly use?
- what do you actually spend your time doing?
- do these pieces of kit make your fingers cold/numb?

After quite a lot of beating around the bush you eventually get something of an occupational history from Kevin. He spent his youth 'on the roads' going from one motorway repair contract to another. When he married and settled down he worked in metal shops and foundries in semi-skilled work.

During this stage of your discussion, it emerges that Kevin is not only anxious about the possibility of multiple sclerosis but is also worried that the lack of feeling in his fingers is making him a bit clumsy in his work and so he is having problems achieving the piece rates which improve his take-home pay. He wonders too if the occasional buzzing in his ears may be connected to all this.

Problems you might anticipate in referring Kevin for investigation

Although HAVS is quite a common condition it is not terribly well known and it may prove difficult in your area to refer him to a specialist who is aware of the condition and able to assess it effectively. Ask Kevin if his employer has an occupational health or safety function. If they have, Kevin should be encouraged to report to it and be investigated by that route. If not, your local NHS Trust occupational health service may be another point of contact; a proportion of these now run general occupational medicine clinics and, even if they do not, they should know who amongst local consultants has a special interest or knowledge of HAVS.

The consultant's letter concerning Kevin comes back as follows:

Dear Dr

Re Kevin R, etc.

This man has severe (Stage 3) HAVS with Stage 3 sensorineural components. I have advised him of the prognostic implications of this. It is clear that he will need to change his work. His employer will need to notify the condition.

Yours sincerely,

The prognosis of the condition and staging system

The first staging system for HAVS, or vibration white finger (VWF) as it was then known, was a UK system based on subjective, perceptual criteria such as tingling, numbness, blanching severity, etc. It has been superseded by the Stockholm Classification which divides the vascular and sensorineural components. Stage 3 vascular symptoms are severe, frequent attacks affecting all phalanges of most fingers. Stage 3 sensorineural symptoms include persistent numbness with tingling, reduced sensory perception and reduced tactile discrimination and manipulative dexterity.

The prognosis is poor. Only 20–30% of those with earlier stages of the condition recover on removal from exposure; recovery from the later stages, as in Kevin's case, is rare. Management is symptomatic and essentially avoidance of further vibration work.

Much work has been done on controlling tool vibration at source and this can be complemented by work cycling, personal protective equipment (PPE) and training of workers about HAVS risks. There are *now* specific diagnostic tests for HAVS; the immersion test described in the case history is not specific or absolutely reliable but is 'do-able' in the surgery.

Notification systems for occupational diseases

A list of diseases, of which HAVS is one, has to be reported to the Health and Safety Executive by employers under RIDDOR regulations. A similar list of prescribed diseases makes people eligible for Industrial Injury Benefit. Reporting under these schemes greatly underestimates the burden of occupational disease in the UK. Therefore, HSE has made other initiatives including reporting schemes for respiratory diseases (SWORD), skin disease (EPI-DERM) and other diseases (OPRA). These combined intelligence systems are known as ODIN (Occupational Disease Information Network). Another population-based approach is via the periodic UK Labour Force Surveys (work-related illnesses).

Other occupational problems which Kevin seems to be developing

Noise-induced deafness. Kevin appears to have some tinnitus and early social deafness. This is unsurprising as all his jobs have been noisy. Typically the condition takes decades to appear so for the early symptoms to present at Kevin's age would be as expected.

Ear, nose and throat departments are experienced at dealing with noise-induced hearing loss. He will be encouraged to use personal protective equipment; ear plugs and muffs are very effective. The employer is required by law to assess significant risks from noise and to introduce control measures to reduce occupational exposure. It should be possible to preserve Kevin's hearing at its present level if the reasonable precautions as detailed above, are taken. A GP might also wish to ask Kevin whether he or she could approach his employer concerning the need to notify the condition and to assess exposure to noise.

Further reading

Health and Safety Executive (1994) *Hand–Arm Vibration.* HSG 88. HSE Books, Sudbury.

Pelmear PL (1999) Vibration (hand–arm and whole body). In Baxter PJ *et al.* (eds) *Hunter's Diseases of Occupations* (9e). Arnold, London.

Coggon D, Palmer *et al.* (1999) *Hand-transmitted Vibration: occupational exposures and their health effects in GB.* Contract Research Report No 232. Health and Safety Executive, London.

Scenario 3: dermatitis

Meena is a 27-year-old nurse. The skin of her fingers, particularly the sides, shows reddening, vesiculation, roughening and scaling. She says that she has had several episodes of this over the last couple of years which she has self-treated with emollients or steroid creams that she has 'borrowed' from colleagues.

Initial questions to elucidate what is going on here

This should be getting very familiar!:

- what is your work?
- what does it actually involve?
- what substances do you work with?
- do you know the hazards of these substances?

and in this case:

- do you have to cleanse your hands frequently?
- do you need to wear gloves regularly to do your job?

When you go into her past history, Meena tells you that the trouble usually comes on when she is doing a lot of wet work, even though she always uses gloves for this. Before becoming a nurse, she worked in a food factory for a couple of years. She had the same trouble there on a number of occasions and the factory nurse moved her from one job to another on several occasions. In the end, Meena said 'I got skin trouble wherever I went, so they had to get rid of me, but they did give me some money'.

Implications of the past history for the present condition and its management

It looks as if Meena is prone to getting dermatitis when she uses her hands a lot, particularly if her natural protection is broken down by regular and frequent washing. This implies that she will be at risk in any working environment which involves such activities.

The cycle that she went through in the food factory is rather typical of what happens in establishments where there is risk of dermatitis (the same is true where there is a risk of work-related upper limb disorders, *see* Scenario 7). Often, the procedure is agreed between employer and trade unions. If the patient gets dermatitis in a particular job then the usual tactics of avoidance, substitution, protection and treatment are applied. If nevertheless the condition persists another job is tried (back to avoidance). Dismissal, with or without some cash settlement, may follow after three or four unsuccessful moves.

Other implications are that she may be at 'special risk' and therefore in legal terms the employer may owe Meena a duty of 'special care'. This would be discharged by training, close attention to her system of work and the requirement to report any problems at an early stage.

Referral and investigation

The first port of call will be the Trust's occupational health unit (they nearly all have one now). They may wish to refer her on to a dermatologist or her GP may want to do this. Quite often the problem and its solution are quite clear without needing to do this.

▼

Most dermatologists will arrange to have patch testing done to the standard battery of industrial agents which of course can be supplemented by tests of extracts to specific agents under suspicion. This is quite often done in a targeted 'contact dermatitis' clinic.

Following referral, you receive the following letter:

Dear Dr

Re Meena P. age 27, etc.

Patch testing on this patient of yours produced the following results of interest from the standard battery of industrial agents:

aldehydes	+ve
isothiazolinones	+ve
rubber chemicals/latex	−ve

I have treated your patient and advised her accordingly.

Yours sincerely,

Advice on prevention of exposure, management and avoidance of contributory factors

It looks as if Meena works with gluteraldehyde and has developed allergic contact dermatitis to it. The patch tests also show that she is sensitive to isothiazolinones which are constituents of bactericidal washes often used in the food industry. Fortunately she is not sensitive to rubber chemical or latex or she would have real problems wearing gloves at all.

If she is currently in exposure to glutaraldehyde (and her history suggests she may well be) then the cycle of management approaches discussed earlier in this section comes into play. Which is deployed when, is a matter of individual judgement consulting the wishes of the patient. This is hard to do without close contact with the hospital OH department.

Meena's history is strongly suggestive of an irritant component aggravating the condition (the commonest contributory factor). This is a situation where dose (cleansing) can be titrated against effect (irritation) by limiting the number of 'cleansings' to the minimum. Very regular use of moisturising and texturing creams helps a lot practically in addition to the short courses of steroid creams which may be needed for the eczema. More complex techniques using occlusive treatment, cotton-lined gloves and other adjuncts may be kept in reserve.

Further reading

Rycroft RJG (1999) Occupational diseases of the skin. In Baxter PJ *et al.* (eds) *Hunter's Diseases of Occupations* (9e). Arnold, London.

Scenario 4: dread exposures and dread diseases

Mavis and her older sister Aggie, who are 68 and 75 years old, respectively, have moved to your patch quite recently to be near their children who have settled in the area. One morning, Mavis presents at the surgery with a three months' history of loss of weight, cough and intermittently bloody sputum. In her characteristically blunt way she assumes that she has lung cancer; as she says 'I were a smoker since I could get me nose over t'counter'.

On referral the diagnosis is confirmed as a primary carcinoma of the bronchus which is inoperable with both local and some distant spread. Palliative treatment is quickly effective in giving Mavis a decent quality of life. She comes in monthly to get her pain relief adjusted and, on one of these occasions, mentions that both she and sister Aggie had worked in their youth in a factory in Yorkshire making brake-linings. She recalls that there was some fuss made by her union about asbestos but 'it didn't come to nowt'.

A few months later her older sister Aggie presents with loss of weight, cough, shortness of breath and chest tightness. She has also been a lifelong smoker. In her case, a diagnosis of mesothelioma is rapidly made and she succumbs to it soon afterwards.

You are visited by Daniel, Mavis' grandson, during the course of the illnesses experienced by Mavis and Aggie. He is a newly qualified lawyer and is considering taking action against Mavis and Aggie's former employers. Having received written consent from the remaining sister, he has a series of questions for you which address the following topics.

The likelihood that Mavis and Aggie's diseases are work-related

Mesothelioma of the pleura is regarded as almost pathognomonic of asbestos exposure. However, that exposure may not be occupational but environmental in the sense of having lived in the vicinity of an asbestos works or having had relatives who worked at such places and came home with grossly contaminated clothing. Establishing the work-relatedness of lung cancer is much more problematic although there is a tendency, especially judicially, to attribute it to asbestos exposure where this has been shown to be present historically. This is because of the effect of the interaction between asbestos and cigarette smoke in initiating and promoting lung cancer.

Steps in making a causal connection

The Industrial Injury Benefit Scheme is underused and, in general, your patients should be encouraged to apply if you have reasonable suspicion of an occupational cause. The steps which are used to ascertain eligibility of industrial injury benefit from a prescribed disease (*see* Scenario 2) may be seen as a model for attributing causation. It is necessary to establish that exposure took place usually by ascertaining that the individual worked in the appropriate industry and in the appropriate job(s). This may be backed up with data on historic dust or other exposure levels (some records in the asbestos industry now go back decades). Also the disease must be present and must be one which has been shown epidemiologically to be present in significant excess in workers in a particular industry. Other factors which may come into account in litigation are experimental evidence in animals (toxicology) and appropriate latency. Latency for mesothelioma is 30–40 years and this is not untypical of other industrial carcinogens.

Thus the inference of causation in occupational disease is a complex and often arguable matter whether for 'dread disease' or other more mundane conditions. Many medical practitioners appear over-willing to infer occupational, and indeed environmental causations, whilst unaware of the scientific basis of dispute over specific exposures and specific conditions.

Further reading

Harrington JM *et al.* (1999) Occupational cancer, clinical and epidemiological aspects. In Baxter PJ *et al.* (eds) *Hunter's Diseases of Occupations* (9e). Arnold, London.

Scenario 5: stress at work

George is 34 years old. He is a police inspector and a rising star in the local constabulary, being frequently in the local media in association with the latest, new policing initiative.

His wife Mary voices her anxieties about George and his behaviour over the last 18 months or so, on one of her visits to you for a contraception check-up. She says that he works very long hours, has become irritable and is bringing more and more work home. She is worried that his relationships with her and the children are deteriorating. You suggest to Mary that she mentions her conversation with you to George and you ask her to invite him to make an appointment to see you.

Knowing something of George through his public persona and some limited social contacts, you are somewhat surprised to find that he has needed little or no persuading to come to see you. The very next week he has made an appointment and is sitting in your surgery looking very subdued and hunched up in the chair. It needs only a few words from you to release the 'log-jam'.

There are problems in every compartment of George's life. He can't concentrate, he feels that people are picking on him, both at home and at work, and that no-one respects him. His career had been going well but has been stalled by a combination of organisational 'freezes' and failure at the last two promotion boards.

There is little to find physically when you examine him, although you note from your records that he has put on a couple of stone in as many years. He responds to your probing questions in an equable way, with some quiet humour. He is somewhat bitter about the situation at work and says that 'hard choices and decisions are not taken because senior officers prefer to take soft and politically correct options'. He is quite philosophical about the promotion boards saying that he realises he has a few hard edges that need grinding off.

On questioning he tells you that he takes far less exercise than he did two or three years ago and the family social life is curtailed compared to how it used to be. He admits to dwelling on an incident that occurred when he was a young constable; he had rescued a child who had gone through the ice on a frozen pond and subsequently received an official commendation. In retrospect he had realised that he very nearly died in the incident himself and that what he did was actually very imprudent.

The initial differential diagnosis is:

- depression
- anxiety
- stress
- psychotic disease
- a combination of the above.

Pressure and stress in the occupational context

Stress is a very loose term, although its very vagueness makes it a useful social concept. 'I've been under a lot of stress lately' may be used to mean 'I've been working hard lately' or 'I've suffered a series of catastrophic life-events recently and I'm profoundly depressed'.

Occupationally, it is customary to differentiate pressure and stress. Each of us has a comfort zone of pressure in which we can work well. Outside this zone there is stress which is thus defined as the unacceptable pressures of work. This model can be represented diagrammatically thus:

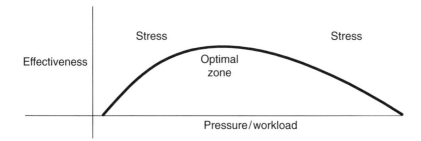

Figure 6.1

In this model both too much and too little pressure can create stress. A number of psychological instruments such as the General Health Question-naire (GHQ) and the Occupational Stress Indicator (OSI) measure stress, coping, coping mechanisms and other parameters of stress. It is argued that sufficient is known about the subject to make practical adjustments to the workplace in order to relieve it.

Susceptibility of professionals and senior managers to stress compared with other workers

There is an increasing perception that professionals and senior managers are more likely to be resistant to stress. There are a number of reasons for this. They have more autonomy and control in their jobs, they are more likely to think systematically and prioritise efficiently, they are better at coping and, senior managers at least, are, by definition, survivors.

Coping with stress

It is claimed that people can be trained to understand, anticipate, perceive and cope with stress. The scientific base of these claims, in terms of controlled intervention studies of sufficient duration, is sufficiently scant for the claims to be somewhat optimistic at present. However, a number of employers are introducing training programmes (sometimes called 'stress inoculation'). Employee Assistance Programmes (EAPs) usually offer confidential coun-sellors working outside the employer/employee framework. In practice only about 30% of EAP uptake is associated with work problems. The objective benefit of EAPs remains largely unevaluated.

You decide that George is depressed and that he is suffering from some stress. You prescribe antidepressants and refer him to the community psychiatric nurse. You arrange to see George a month later.

When he comes to see you again, George says that he did not take any of the pills because 'he didn't want to get addicted' and anyway he wanted to 'give the nurse a try'. He says that the two counselling sessions he has had so far have been very supportive and useful and that he feels there is a lot more coming out than he expected. He says he has thought about things a lot and can now see no future for himself in the police force. He plans to resign and sue his former employers for stress. He makes it clear that he expects you to support his plans.

Your options

Most larger organisations will not wish to lose people such as George and will be surprisingly understanding and supportive of people in crisis. All the major emergency functions, including the military, are moving towards dealing effectively with, and retrieving people from, stress in general, and post-traumatic stress in particular.

The GP or community psychiatric nurse should make contact with the occupational health unit attached to George's force, with his written and informed consent. If an individual is away from his workplace on sick leave, the likelihood of his getting back into the workplace is greatly diminished by the sense of isolation and the lack of value that absence from work engenders. A better option which is usually feasible, is a quick assessment of the key stressors, followed by relocation by a temporary arrangement to a situation where those stressors are absent or diminished. This buys time to explore the deeper issues underlying the crisis, including the trauma.

Further reading

Cox T (1993) *Stress Research and Stress Management: putting theory to work.* CRR61. HSE Books, Sudbury.

Scenario 6: acute and chronic intoxication

Winston is an old patient of yours. He is 58 years old, and a hypertensive with ischaemic heart disease. He has mild angina. His compliance with his treatment regime is spasmodic.

Despite all this he has held down a physically demanding job in a metal-finishing workshop for many years. It is quite a big local enterprise which does plating, coating and painting.

One day you receive a discharge letter from the local casualty department concerning Winston. He was admitted overnight for observation after what was described as an acute intoxication following a solvent spillage at work. Winston comes in the next day seeking reassurance.

Initial evaluation

You will be an old hand at this now:

- what happened?
- what were you doing?
- what was/were the substance(s)?

Further advice and information

This sort of incident has to be reported to the HSE. Call the local HSE office and speak to one of the doctors about the likely effects of the spillage. Your local hospital OH department may be able to give advice. You might contact the National Poisons Service by phone and search on-line services such as Medline, Toxline and chemical company websites.

Prognosis of the incident

Do not give advice until you are sure what the substance was and what its effects might be. Often these depend on concentration, duration of exposure and other factors of which you will be unaware. If you have any doubts it may be prudent to refer Winston to an occupational health expert. Most solvents will not have prolonged effects from a single exposure.

Six months later, Winston's wife Mary, normally a very calm and matronly lady comes to see you. She says that their son John came home for a visit recently and remarked to her that 'Dad seemed very knocked off'. She has been wondering about this and thinks that he has got slower and less interested in things generally over the last few years. She asks you if this might be due to the chemicals at work and might it have got worse since the spillage.

Chronic effects linked to the workplace

There has been controversy whether relatively low-level solvent exposures maintained chronically (but below hygiene standard control limits) can cause subtle dementia-like neurological effects. Such a syndrome has been diagnosed in Scandinavia, and as such, is often referred to as 'Danish painter's syndrome'. The criteria used to diagnose these conditions have been widely criticised as being highly subjective and the controversy continues.

To demonstrate that Winston's deterioration was due to solvent exposure as opposed to ageing, stroke, alcoholism, etc. is difficult. Only a formal study of Winston's workplace looking at exposures and the histories of those who have worked there could possibly answer the workplace linkage question. A topic without simple answers!

Further reading

Levy LS *et al.* (1999) Aliphatic chemicals. Aromatic chemicals. In Baxter PJ *et al.* (eds) *Hunter's Diseases of Occupations* (9e). Arnold, London.

Scenario 7: Repetitive strain injury and work-related upper limb disorder (RSI/WRULD)

Claire and Anna are both patients of yours who work for a large local company as data entry operators. They come to see you, both complaining of upper limb symptoms which are mainly a problem at work.

Claire is 24 years old, recently married, and her symptoms are of aching and shooting pains in the back of the hands, where her right hand is worse than her left. Sometimes the pains radiate to the right forearm and shoulder. On questioning, the symptoms appear to be intermittent but described as definitely getting worse at work. Claire says that 'there is a lot of this RSI at work and management don't do anything about it'. On examination, you can find nothing except slight tenderness of the right common extensor origin.

Anna is 50 years old and has worked for the company for many years. Until recently she was the chairman's secretary but in the takeover which has just been completed, the chairman's post disappeared, as did her job. Her symptoms are similar to those of Claire but are more obviously right-handed and more proximal than distal; in addition, her neck hurts quite a lot. When you examine Anna, her neck movements are rather limited and a brachial stretch on the right elicits the symptoms of which she complains.

Claire asks you to send a letter to her management asking that her work be modified to ease her symptoms. You rather reluctantly agree to do this. You refer Claire and Anna to a local rheumatologist who you know is interested in these sorts of problems. Your reasons for doing this are the possibility of some sort of 'RSI epidemic' in the company where the ladies work and the vehemence of the complaints made to you by Claire and Anna. The following correspondence ensues:

From yourself to the personnel manager, X company

> Dear Madam, In confidence
>
> **re Claire W., etc.**
>
> This patient of mine who works for you, has presented at the surgery with upper limb problems that are strongly associated with work. She has asked that I write to you about this.
>
> I have referred her to a rheumatologist who has a special interest in these matters. In the meanwhile, would it be possible to change her work so that she is not continuously working at a keyboard? This should help to keep her symptoms in check.
>
> Yours sincerely,

From personnel manager, X company, to yourself

> Dear Dr Y   Confidential
>
> **re Mrs C.W., etc.**
>
> Thank you for your letter concerning Mrs C.W. The company takes its health and safety responsibilities very seriously. We are in full compliance with the DSE Regulations, which in the case of Mrs C.W. require periodic assessment of her workstation, eye checks, etc. In fact Mrs C.W. has been involved in applying this procedure as a workstation assessor and has judged her workstation and work arrangements as being in full compliance with the Regulations. (Please see the copy of her assessment attached.) Additionally, we already ensure that staff have regular breaks from computer screenwork by doing filing, sorting, etc.
>
> You may wish to note that Mrs C.W. has had 12 single days off work in the last 10 months. These have all been self-certificated and the reasons given have never been anything to do with her hands or arms. You will appreciate that this level of sickness absence usually triggers investigation which may lead to disciplinary action.
>
> We consider that the company is doing everything that it can to help Mrs C.W. come to work regularly and be comfortable in her job.
>
> Yours sincerely,

From the rheumatologist to yourself

Dear Dr Y

re Claire W., etc.

Thank you for referring this interesting patient. Clearly she has RSI and I think she should be absented from work for some time since the condition is being so aggravated by it. I have arranged for her to start physiotherapy and will see her again in a month's time.

It does look as if there is an outbreak of RSI at X company, which mirrors the events described in the Australian literature of the early 1980s. I have written to the company, with Mrs W.'s permission, to suggest a full survey of the workforce. Thank you for drawing it to my attention.

Yours sincerely,

Also from the rheumatologist to yourself

Dear Dr Y

re Anna P., etc.

Thank you for referring this patient. X-ray examination reveals that she has quite well-advanced, diffuse, cervical spondylosis with some nerve root irritation. There is not much one can do about this except to keep the neck as mobile as possible. Mrs P. feels the loss of her status as the former chairman's secretary keenly, and it may well be that she should use this current problem as the basis for retirement on grounds of ill-health.

I do not doubt that this would develop into RSI if it is not dealt with now. The wider amplitude and more frequent head movements needed for keyboard and screen work will undoubtedly be an increasing problem for her. I have suggested to Mrs P. that she come and discuss how to proceed with your good self.

Yours sincerely,

And a bit later to yourself

Dear Dr Y

RSI epidemic at X company

I have received the attached letter from X company. I think it is really rather offensive and that we should do something about it. I would welcome your views.

Yours sincerely,

Enclosure (from X company to the rheumatologist Z)

Dear Dr Z

RSI

We have received your recent letter concerning this subject in which you express the wish to carry out a survey of 'RSI' at our company on the basis that there appears to be an 'epidemic' of the condition occurring in the organisation.

We are unsure why you have developed these views since you appear to have seen only two patients who work at company X. For the record, our nursing staff have checked back through their surgery notes for the last five years. These show a very low level of problems with upper limb disorders and no recent increase in reported problems. Your letter was shown to our group occupational physician who has written as follows:

'RSI is an unhelpful diagnosis, wrong in its inference of causation (repetitive), wrong in its description of what has happened (injury), and wrong in its attribution of pathology (strain). The trouble with the Australian experience is that most doctors have read the papers that came out at the start but not those recanting papers that came out at the end of the "epidemic". So we are left with groups of professionals who believe uncritically in the concept of "RSI".'

We thank you for your interest in this matter and assure you that we have the matter fully in hand in medical terms and in relation to our compliance with the law.

Yours sincerely,

Later: you get from Dr Z a copy of another letter sent to him. Letter from Group occupational physician X company to Dr Z, rheumatologist

---

Dear Dr Z

I gather that our local company wrote to you recently quoting a memo that I had sent them about RSI. I'm sorry this got out to you in the form that it did, since the memo was intended only for internal information and the quotation that you were sent was an abbreviated version of what I actually wrote. Nevertheless what I said is true and I attach some key papers in amplification for your information.

The local management and nurse at X company do take their responsibilities under the DSE Regulations very seriously as indeed they do for any problems reported by employees. One of the ladies whom you saw helps her new husband with his carpet-laying business and this may have something to do with her symptoms and the recent deterioration in her short-term sickness absence record.

Yours sincerely,

History of the conditions, work-related upper limb disorder (WRULD) and repetitive strain injury (RSI)

RSI was a phrase coined during the early stages of the Australian WRULD epidemic of 1981–87. The phrase has since gained world-wide currency. The epidemic was seen in 'white-collar' workers and created enormous medical and media interest. At an early phase in development of the epidemic an official medical committee described the condition as a five- or seven-stage process leading to chronicity. Typically symptoms were diffuse, flitting and with few signs of note. The conditions did not respond well to rest, splintage, physiotherapy, etc., and persisted long after such 'blue collar' conditions as tenosynovitis would have been relieved. Generally, the epidemic was attributed at the time to the introduction of new technology, but later careful examination of contemporaneity showed that this was not a plausible hypothesis. Other theories put forward have been iatrogenesis, psychosomatic disease and even mass hysteria!

The epidemic died down in 1985–87; the reasons are controversial. Some authors cite a robust official debunking of the condition, others the failure of a highly publicised test case in the Australian courts and yet others the public contradiction of their previous position by some members of the medical committee. But by then the idea of RSI and its putative aetiology had entered the medical literature and popular culture world-wide.

WRULD is a much vaguer term and encompasses RSI-type problems as well as the very definite and specific syndromes such as tenosynovitis which are often found in manual workers.

Certain groups of health professionals (who might be expected to give medicolegal opinions) appear to 'believe' in the existence of RSI (e.g. physiotherapists, rheumatologists) and others (occupational physicians, orthopaedic surgeons) do not. Some professional medical groups perceive work as something to be avoided when unwell whereas others see it as important to help people to stay at work whenever possible because of the central importance of work as a social (and economic) determinant. It is unlikely that these different perceptions will diminish in the short term and it is thus important to acknowledge that they exist and understand why.

Until recently there has been something of a societal tendency to view blue-collar WRULDS such as tenosynovitis as an inherent concomitant of the work (much the same as was noise-induced hearing loss, one or two generations ago).

Recent research has identified a number of risk factors such as repetition, excessive force and awkward posture which are clearly avoidable or can be ameliorated by job design and use of appropriate ancillary equipment. There are now legal requirements for employers to assess and reduce workplace risks, including those of upper limb disorders and risks from manual handling.

This also applies, of course, to jobs where the risk is of neck, back or lower limb injury. These are actually more common than WRULDS.

Ways in which symptoms in case history could be resolved or at least ameliorated

This is nicely and probably accurately encapsulated in Dr Z's letter to Dr Y. Screenwork, especially when transferring to and from paper records does require a wider range of side-to-side and up-and-down neck movements than 'old-fashioned' typing. It can be a genuine 'effect of health upon capacity to work'. An option attractive to some older workers may well be retirement but job modification (e.g. 'little and often') is feasible in practice.

Inferences of situation for GP as advocate

Caution.

Further reading

Health and Safety Executive (1990) *Work-related Upper Limb Disorders: a guide to prevention.* HSG 60. HSE Books, Sudbury.

Society of Occupational Medicine (1997) *Work-related Upper Limb Disorders*. Society of Occupational Medicine, London.

The effect of health on capacity to work

In the first section of this chapter the scenarios were of increasing complexity and described common conditions where work may affect health. Several of these scenarios involved the effects of ill-health in relation to capacity to work. Inevitably, these two basic facets of occupational medicine are often in practice intertwined. We now go on to consider situations where the latter facet, the effect of ill-health, is the dominant factor. These situations can be, and often are, the most difficult.

There are several reasons for this. First, you are arguing from a strong position within your own expertise, the knowledge of your patient and his or her medical position. Second, you are likely to be asked to take some kind of advocacy role. Set against these considerations is that you will be interacting with a system which may be unfamiliar to you, be it a sickness management system, a retirement system or a state benefit system. It is possible that you will find that system unlikeable, implacable or silly but its rules are the ones to which your patient will be subject.

Scenario 8: capacity to work

You have referred your patient Donald, aged 61, for a surgical opinion with symptoms that may be due to gastric ulcer or hiatus hernia and which have not responded well to medical treatment. There is about a five-month wait for an appointment. In the meanwhile, Donald explains that his job as a process operator in a plasterboard factory involves a lot of stretching and bending whilst manoeuvring awkward and heavy boards. This produces a lot of uncomfortable symptoms. You give him a sickness absence certificate and renew it regularly.

A couple of months later, you receive an irate call from the works manager at Donald's firm. He tells you that Donald's foreman had called at Donald's house with some gifts and cards from work. He found scaffolding up round the house and Donald was half-way through painting the outside of the house. The suggestion put to you is that if Donald can paint his own house then he can do the same job at work and that they are very willing to find such work.

The response

The author of this chapter has been at the receiving end of responses from GPs in his post as an occupational health physician that have ranged from 'oh dear' to 'get lost'. A reasonable middle way is to recognise that the GP gives a patient advice in these circumstances and that advice is necessarily based on what the patient tells him or her.

A couple of days later it is Donald's turn to be irate. He appears at the surgery with a letter from the works advising him that he has been dismissed for gross misconduct; apparently he had already received a final warning for various misdemeanours. Donald says that he was only up the ladder on one occasion, helping his brother who was actually doing the work – the foreman just happened to come at the 'wrong moment'. Anyway it was good for him to be out in the sunshine a bit, etc.

Donald asks you to write to his former employer to emphasise the genuineness of his medical condition. The following correspondence ensues:

Yourself to company A

> Dear Sirs
>
> **re Donald B., etc.**
>
> My patient, Donald B., has asked me to write to you in support of his claim for reinstatement. He has been a patient of mine for some years and I am quite satisfied that his symptoms are real and troublesome. It would be humane to reconsider your somewhat harsh and arbitrary decision.
>
> Yours sincerely,

Company A to yourself

> Dear Dr C
>
> **re Donald B., etc.**
>
> Thank you for your recent letter. Our decision with regard to Mr B. was neither harsh nor arbitrary. We have a disciplinary procedure, agreed with our Unions and compliant with UK employment law. It is a measured process capable of appeal and requiring due procedure at each stage. Mr B. was at the final stage of this process prior to this last episode in which you have become involved.
>
> We have no particular reason to doubt the genuineness of Mr B.'s symptoms. Had you written to us suggesting alternative work which avoided

those factors which aggravated Mr B.'s symptoms, we would have made every attempt to modify his work. Certainly we could have found him work identical to that which he was found doing around his house. The truth of the matter is that we are concerned here not with illness or symptoms but with capacity to work. Mr B. was capable of work and despite your advice to the contrary, chose to do so but not at our premises.

With regard to your prior telephone conversation with Mr D., works manager, we do of course understand that the sick note that you give comprises advice only to your patient. Whilst not obliged to do so, we are normally happy to abide by that advice as if it were given to us but the present circumstances discourage us from doing so in the future.

Yours sincerely,

The position of the general practitioner's sick note

Given that this has been the situation for some years now, the sick note still retains far more official credence (with State and employers) than its status should realistically command. If doctors have ceded their traditional gate-keeper role in society then truly neither State nor employer should pay any regard to the sick note. Probably we are in a transition to a situation where they will not.

A GP's current obligations are to:

- provide advice to patients on whether they should refrain from their usual work
- only record this advice on an official form (Med 3) when it is based on an assessment of the patient's medical condition or disability
- consider alternative management options such as, 'fit for alternative/ light duties' or referral to the specialist Disability Services of the Employment Service if this is in the best interests of the patient.

But GPs are not required to certify long-term incapacity or incapacity for any work. The IB204 Guide has more information on these requirements.[1]

Other methods that might be available for certification of absence attributable to sickness should sickness or incapacity be the subject of the certification

Perhaps the most honest system would be to acknowledge reality, which is that it is the individual him/herself who decides whether or not to absent themselves from work. The Dutch had a system (not well liked) where only special social security doctors could certify ill-health eligible for benefits. All systems, including our own, tend to instability because of the subjective nature of the judgements involved.

A way out of this may be to look not at the illness but the incapacity that it actually causes – a change in emphasis that is currently being developed by the UK Benefits Agency (*see* Scenario 10).

Scenario 9: 'light work'

Gwen is 36 years old and a mother of four children; she works as a packer at a food factory. She is fat, unfit and you know that her husband is a rather unreliable man who is seldom in employment. Gwen has had a 'bad back' for some years which gets particularly troublesome round about the school holidays or at least that's how it seems when you look at the practice notes. There are never really any localising signs and the problems seldom last longer than a week or two. She has had a bit of physiotherapy on a few occasions mainly to encourage her to exercise to strengthen her back but she seldom turns up more than once for this.

Now, she comes to the surgery complaining of her back again and on examination she certainly appears to have a problem. The signs are rather non-specific with flattening of the lumbar lordosis, some paravertebral muscular tenderness at L3–S1 vertebrae on the right and straight leg raising is a bit limited on the right. She asks you to write a note suggesting that she be permitted to do 'light work' for some weeks whilst her back settles.

Risks to your patient's interest in acceding to this request

The employer may choose to dismiss her for 'frustration of contract' on grounds of medical capacity, or rather incapacity. This is actually quite easy to do but is seldom done in practice (currently) by more reputable employers. The other main factor in practice is that 'light job' often means 'light pay-packet'.

Fortunately the response from the factory OH department to your note is positive:

Dear Dr

re Gwen H., etc.

Thanks for your recent note. We have found Gwen work in the label control department where her extensive experience of the Company's product lines will be invaluable. There is no heavy lifting in this job at all.

> The walking and climbing involved in the job should help Gwen's fitness, which is none too good.
>
> Sister, OH Department, Factory F.

All seems to be well for a while but then Gwen calls again. Her back is better and she wants to go back to her old job. She complains that there is no overtime in her new job and that because of the 'quality control' aspects of her new job she is no longer seen as one of the girls. You write another note. The response this time is not so sympathetic.

> Dear Dr
>
> **re Gwen H., etc.**
>
> I was a bit surprised to get your recent note. Gwen was told at the time of transfer to her job in label control that the move was permanent. She seemed very happy to make the move, particularly as the rate of pay was higher. It is true that just now, round about Christmas, there is quite a lot of overtime in the sort of job she used to do, but again she was warned about this and anyway there is a possibility that, with her back history, overtime would have been ruled out as a precautionary measure.
>
> Sister, OH Department, Factory F.

Your response

This sort of situation is not uncommon, with an element of the 'grass being greener on the other side'. All employers, even large ones, have only a certain amount of flexibility and that is diminishing as companies drive for greater efficiency, multi-skilling, etc. Gwen may have pushed her luck to its limit.

If you should want to 'take the Company on', what are you actually taking on – a medical issue or perceived unfairness? If it is the latter, on what authority do you act?

Scenario 10: ill-health retirement

Norman, a patient of yours, develops bowel cancer at the age of 53. The prognosis on early assessment is good although he will have to undergo a fairly prolonged and grinding course of treatment. His general physical state is good but he gets tired easily and certainly isn't up to his rather demanding job as a regional sales manager. He makes enquiries with his employer about taking ill-health retirement (IHR) and is advised that he is likely to get sympathetic consideration particularly if you, as his GP, will support a parallel claim for state incapacity benefits.

A letter arrives for you confirming these possibilities. It is from the employer's medical adviser. It confirms the advantageous terms for which Norman is eligible and sets out the terms of the company scheme (which requires permanent disability as a criterion for entry). You write to advise this medical referee that you consider it highly likely that Norman will never really be fit to do his very onerous job.

Your preliminary discussion with Norman

Norman may well be feeling pretty grim. With the best will in the world, the future horizon may not be all that far ahead as he sees it. He will probably view IHR as a blessed option that relieves him of worries in the present and offers the prospect of comfort in the future. What he will actually do in that future will be far from his mind.

Given that the prognosis is good and that the man is likely to be conscious of his job status and of a Type A personality the prospect of a return to work and even maintaining contact with work during the illness should be considered. A part-time training or monitoring role might easily be the right thing.

Norman accepts the IHR terms and whilst consulting you to check his treatment, etc., he often mentions what a weight it is off his mind to be able to retire so well provided for. After a year or so he completes his treatment, regains weight and takes up golf and all his other old hobbies with his old characteristic energy and enthusiasm. A bottle of fine brandy arrives at Christmas to thank you for your help and to celebrate Norman's good fortune.

One day, you find him at the surgery waiting for it to open. He tells you that he was called in to the Benefits Agency for a medical assessment. He was assessed as being capable of work and his state benefit has been stopped. It is a relatively small, but still significant part of his income, and he is hopping mad to have it cut off. He wants you to do something about it.

Your response

There are two technical points to make. First, the Benefits Agency procedures now known as Personal Capability Assessment, consider the functional limitations and restrictions which result from the disease or injury rather than the medical condition itself. The assessment covers the ability to perform day-to-day activities and is not concerned with capacity to do a specific job. By law the assessment cannot take any account of factors such as age, skills, experience, attitude or temperament. Second, there is an appeal process, and your comments can be incorporated into this.

Even worse is to follow. He receives a letter from his former employers. They explain that his IHR was based on the likelihood that he was unlikely to do work similar or identical to that which he had previously done. The insurers who provided a large proportion of the pension that he was receiving also require him to undergo medical review.

When this is done, the verdict is that he is fit to resume his former work. Somewhat apologetically, his former employers offer Norman his old job back.

Reality of this scenario

Beware – scenarios like Norman's will become commoner! The rules of various schemes are written very differently. Much more arguable is how rules are interpreted. The increasing fragmentation of large pension funds (as organisations 'downsize', 'demerge', etc.) and their stricter actuarial reckoning is meeting up against the vast likely increase in demand from the greying of Western populations. Result – tighter rules and tighter interpretation as a general and lasting trend for the future.

Further reading

Seaton A, Agius R, McCloy E and D'Auria D (1994) *Practical Occupational Medicine*. Arnold, London.

Reference

1   Department of Social Security (2000) *Guidance for Doctors. 1B204 Guide*. Department of Social Security, London.

Reflection exercises

If you want to consolidate your learning, you need to spend some time thinking about how what you have read applies to your own practice. Why not do one, or both, of the following exercises?

Exercise 1. Quiz about relationship of occupation and cancer (*see* Scenario 4 – dread exposures, dread diseases)

(Tick all boxes that you consider represent a true statement – one or more will be correct.)

1 Occupation is the main causal factor in the following percentage of fatal cancer cases:

   <1% ❑   20% ❑   5% ❑   >25% ❑

2 In relation to occupational and environmental factors there has been an epidemic of cancer since:

   1850 ❑   1900 ❑   1945 ❑   never ❑

3 The following are *common* causes of cancer:
   Occupational exposure to:
   radiation   ❑
   asbestos   ❑
   dyestuffs   ❑
   benzene   ❑

4 Smoking affects the outcome of occupational cancer as follows:
   lung cancer due to asbestos, multiplicatively   ❑
   lung cancer due to asbestos, additively   ❑
   mesothelioma, not at all   ❑
   chemical-induced bladder cancer, additively   ❑

5 Current controversies in occupational cancer centre around the possibility of:
   cancer of the liver in sewermen   ❑
   leukaemia in power workers   ❑
   brain cancer in power workers   ❑
   cancer of the uterus in dry-cleaning workers   ❑

You can find out the correct answers by registering on the distance learning programme or from:

Doll R and Peto R (1981) The causes of cancer: qualitative estimates of avoidable risks of cancer in the US today. *J Natl Cancer Inst.* **66**: 1193–308.

This Exercise might take 1 hour.

Exercise 2. Analyse three of your patients' records where the presenting conditions were due to the effect of work on health. Use the learning points from this chapter in your analyses.

Take three examples of different situations or conditions, which you can recall or take the next three consecutive patients who present with symptoms and signs that are likely to have arisen as a direct result of their work.
 Describe:

- the symptoms and signs
- what their occupation is and what exactly they do at work
- what the relationship is between their work and their health
- whether their work can be changed so that the hazard or threat to their health is minimised or avoided
- whether the effects to their health can be prevented or minimised if their work remains unchanged
- what changes or action you recommend
- predict the future outlook for that patient if (i) no changes are made (ii) the action you recommend is taken up by their employer or (iii) the action you recommend is taken up by the worker him/herself.

This Exercise might take 2 hours.

CHAPTER SEVEN

Occupational health and safety concerns of general practitioners and their staff

The material in this chapter describes the responsibilities of general practitioners as employers and practice managers to themselves, their employees and others involved with the practice. It should help GPs and practice managers to develop the necessary knowledge to discharge their responsibilities in a competent and efficient manner and to be able to handle common occupational health issues in the practice. This should include pro-active and reactive approaches; incorporating risk assessment, compliance with the law, health surveillance and health promotion into the routine of the practice organisation.

Practical guidance is given about fulfilling the duties under the relevant legislation. Various non-statutory aspects are also covered such as managing ill-health amongst practice staff.

Health and safety management

As an employer in general practice you will wish to protect your staff and all other users of your premises from ill-health and injury and have a duty to do so under Health and Safety legislation. Health risk management is about identifying and controlling risks before they cause problems. The initial assessment of risk and subsequent management of health and safety within the practice can be equally successfully performed by a doctor with an interest in occupational health and safety or the practice manager. Effective management will require commitment from all members of the practice team to the principles of safe working practices. The stages are as follows.

1 Find out if you have a problem

- Take a fresh look at your surgery premises area by area, noting any hazards. These may be physical, chemical or hazards created by a work process/activity, e.g. a risk of violence.
- Talk to your staff. Find out if they have any concerns about how their work may affect their health.
- Look at your staff sickness absence records. These may give you a clue as to how work impacts on health, and vice versa.
- Obtain advice from suppliers of substances and materials used in the surgery. Relevant information regarding safe usage and potential health hazards, as well as what to do in case of accidental spillage or ingestion should be contained in suppliers' guidance booklets and material safety data sheets. These should always be retained and kept where they are easily accessible for reference in case of an emergency.

2 Decide what action to take

- Undertake a risk assessment – look at the hazards and decide who might be at risk and how.
- Decide whether existing safety or control measures are adequate – could more be reasonably done to reduce risks still further? If so, identify what steps need to be taken to remove or reduce the risks.
- Decide which health risks should be given priority for action. This will depend on factors such as the magnitude of the risk identified and the ease with which the risk reduction measures can be implemented.
- Record your findings.

3 Take action

Agree the necessary improvements with practice colleagues, identify who will take responsibility for each part. Agree timescales.

4 Check what you have done

After an appropriate time, review whether targets set for improvement have been met. Ensure that staff are complying well with any new safety procedures introduced.

Good management is an ongoing process. You will need to revisit these steps periodically, and when any significant changes are made to your surgery premises or the activities therein. Good health risk management will anticipate problems rather than deal with them after they have occurred.

Case study

An unsupervised child was able to enter a treatment room in a surgery and drink a quantity of carbolic acid (phenol). This led not only to distress and injury for the child and family, but also prosecution under the Health and Safety at Work Act, a substantial fine and a large financial settlement for personal injury for the general practitioners involved.

The Health & Safety at Work, etc. Act 1974 (HSAWA)

Duties of employers

The basis of British Health and Safety Law is the Health and Safety at Work, etc. Act 1974. The general duty of employers under this Act is to ensure, so far as is reasonably practicable, the health, safety and welfare of their employees and themselves, as well as other users of the workplace premises. This will include patients, ancillary staff such as visiting community health workers, contractors and representatives in general practitioners' surgeries. The principle of '*so far as is reasonably practicable*' allows for the degree of risk in a particular job or workplace being balanced against the time, trouble, cost and physical difficulty of taking measures to avoid or reduce the risk.

There is a specific duty within the Health and Safety at Work, etc. Act applying to all businesses employing five or more staff (irrespective of hours of work) to prepare and periodically revise, a written statement of policy. The statement must cover the employer's general policy with regard to the health and safety at work of employees and the organisation and current arrangements for carrying out that policy. Most general practices employ five or more people, and even if the practice has fewer than five staff it is advisable to have a written policy, as it demonstrates an awareness of the importance of health and safety to employees as well as the regulatory bodies. If the policy is updated regularly it will act as a prompt to review any changes that have taken place in the practice arrangements, or in health and safety legislation, since the last review. An example of a written health and safety policy is given below, although the content is not mandatory and can be varied according to circumstances.

General statement of health and safety policy for a general practice – an example

It is the policy of this general practice to provide, maintain and ensure a safe and healthy workplace. Equipment and systems of work will be maintained and reviewed regularly. Such information, training and supervision will be provided as is necessary to enable staff to work safely and without undue risk to their health or the health of others.

Responsibility for the health and safety of all other users of our premises and those who may be affected by our activities is accepted.

This policy will be reviewed annually, and if necessary more frequently, to take account of any changes in work practices or in health and safety legislation.

Allocation of specific duties, and arrangements for implementing this policy are detailed below:

Signed: (partners) Date:

Responsibilities:
(a)   Overall responsibility for health and safety rests with:
                                                             (partner)
(b)   Management and implementation of this policy is the responsibility of:
                                                      (practice manager)
       or in his/her absence (other nominated staff member):

(c)   All employees should be aware of their responsibility under the Health and Safety at Work Act to co-operate with the practice health and safety policy.

General arrangements:
First Aid boxes are sited in the following places:

continued opposite

The following persons are responsible for maintaining
and stocking the First Aid boxes:

All accidents must be recorded in the accident book
which is located in the following place:

The First Aider(s) in our practice is/are:

Fire safety:
Escape routes (insert outline map/diagram showing exit routes to be followed
in case of fire):

These are to be kept clear at all times.

Fire extinguisher locations:

Fire extinguishers will be checked annually by:

Fire alarms will be checked weekly by:

Fire practices will be held quarterly.

Contractors/visitors/patients:
All visitors to our premises should report to the reception desk on arrival.
A strict no-smoking policy applies to all users of our premises.
No visitors will be allowed into areas other than the general waiting area,
adjoining corridors and toilets unless accompanied by a member of staff.

Training:
Appropriate training in health and safety will be given to all staff. The person
responsible for arranging training is:

Hazards:
Risk assessments have been carried out under the following Regulations.
(Copies of the assessments can be obtained from:)
    Management of Health and Safety at Work Regulations 1999
    Workplace (Health, Safety and Welfare) Regulations 1992

continued overleaf

Hazards: (*continued*)

Control of Substances Hazardous to Health Regulations 1999
Health and Safety (Display Screen Equipment) Regulations 1992
Manual Handling Operations Regulations 1992
Health and Safety (First Aid) Regulations 1981
Electricity at Work Regulations 1989

Electrical safety – all portable electrical equipment must be checked by:
before use. Periodic checks of mains electrical equipment will be carried out by:
at regular intervals.

Waste disposal arrangements have been made with:
for disposal of waste (clinical)
and with: for disposal of waste (non-clinical)

Duties of employees and others under the Health and Safety at Work Act 1974 (HSAWA)

Employees too have responsibilities under HSAWA to take reasonable care of their health and safety, and that of others, and to comply with any health and safety measures implemented by the employer. *Other visitors* to the premises have a duty not to interfere with, or misuse anything, provided in the interests of health and safety.

Control of Substances Hazardous to Health Regulations 1999 (COSHH)

These Regulations apply to exposure to substances (including infectious organisms) which may be hazardous to health. The Regulations require assessment of the risks, prevention of exposure, or where this is not reasonably practicable, appropriate control measures to be instituted and periodic review to ensure that the measures are still adequate. In addition there are requirements for adequate instruction and training, and where appropriate for health surveillance of employees exposed to hazardous substances.

The main substances in general practice likely to fall under the Regulations are:

- general cleaning agents, e.g. bleach/polishes
- sterilising and cleansing agents used in treatment rooms
- diagnostic reagents
- gases (e.g. 'anaesthetic gases' in cylinders)
- liquid nitrogen
- latex
- clinical waste and pathological specimens including:
   soiled dressings
   used 'sharps'
   urine and tissue samples
- infections carried by patients, e.g. rubella, TB, hepatitis B
- mercury (in thermometers and sphygmomanometers).

This list is not exclusive. The number of substances in use reflect the size, scope and housekeeping practices of the practice.

It is worth noting that packaged medicines for individual use are not covered by the COSHH Regulations and therefore dispensing practices do not have to assess each and every medicine and tablet on their shelves. Advice on how to perform risk assessments under COSHH can be found in the relevant HSE publication (*see* Bibliography).

Although performing an assessment under COSHH may initially seem a daunting task, it will quickly become apparent that it is not so difficult for the majority of substances used. This is because: (a) The first principle of the Regulations is 'elimination' of substances which do not need to be used – a COSHH assessment is an ideal opportunity for a 'pruning' exercise. For example cleaning materials can be reduced to a handful of substances from a regular supplier. Treatment room cupboards often harbour little-used bottles of medications/applications. The practice policy on sphygmomanometer and thermometer purchasing could be reviewed to consider a long-term aim of converting to a 'mercury-free' practice as equipment is replaced. (b) Most substances have no major hazardous properties and control measures will be straightforward, covering points like storage and action in case of spillage.

Assessment of microbiological risk

One of the most problematical risks to clinical staff and some non-clinical staff such as the cleaners, to be considered under the COSHH Regulations is that of microbiological hazards from body fluids and clinical waste, in particular hepatitis B and HIV. More time should be spent on this aspect of the COSHH assessment than any other.

Measures to reduce or eliminate risk of staff contracting hepatitis B and HIV infection will include:

1   Provision of training for all staff involved in clinical procedures involving bodily fluids.
2   A practice policy on immunisation of staff against hepatitis B infection. Staff will generally fall into three categories:
     (a)   non-clinical except cleaners (no risk) – no need to immunise
     (b)   clinical staff – doctors/nurses and waste handlers/clinical specimen handlers and other staff who may handle sharps or be exposed to accidental needlestick injury through work activities. Immunisation should be offered and recommended[1]
     (c)   clinical staff performing *exposure-prone procedures** which present a risk of infection transmitted from staff to patient. For this group, immunisation and serological proof of immunity or freedom from infectivity are necessary.
These are procedures which involve the staff member's gloved hand being inside the patient's body cavity in the presence of sharp objects such as scalpels or bone spicules. Venepuncture, setting up of i.v. infusions, minor surgical procedures, pelvic examinations and normal childbirth are all excluded.
3   A practice 'sharps' policy which describes the protocol for carrying out standard procedures such as venepuncture and sharps disposal, as well as action to be taken following sharps injuries. An example of such a policy follows.

Practice sharps policy – an example

Sharps are items which may potentially cut or penetrate the skin. In practice these include needles, blades, stitch cutters and glass items such as microscope slides. Correct usage and disposal of sharps are essential to prevent injury and risk of transmission of infectious illnesses (primarily hepatitis B and C, and HIV) both to our patients, our colleagues and ourselves.

Immunisation against hepatitis B infection is strongly recommended for all staff members identified as being at risk of sharps injuries. There is no currently available immunisation against hepatitis C or HIV infection, therefore protection depends on the use of universal precautions which are outlined below.

Universal precautions

•   All staff must use and dispose of sharps in accordance with this policy.
•   The person using the sharp is responsible for its correct disposal.

- The sharps bin should be close at hand when performing procedures to allow immediate disposal without carrying the sharp through the workplace.
- Only designated sharps bins should be used for sharps disposal, both on the practice premises and while on home visits.
- Needles should not be re-sheathed, bent or cut.
- Wherever possible surgical gloves should be worn by the staff member performing procedures involving sharps. They *must* be worn when the staff member has cuts, abrasions or broken skin on the hands.
- Sharps bins should be correctly assembled according to manufacturers' instructions, kept out of reach of children and other vulnerable groups such as confused patients, never filled beyond the maximum fill line and stored safely once sealed and awaiting removal from the premises.

Accidents involving sharps

Action after a healthcare worker has been exposed to blood or other potentially infected body fluids will take account of the interests of both the worker and source patient. The circumstances which led to the exposure will be reviewed and all possible steps taken to prevent a recurrence.

Staff involved in sharps accidents must notify the practice manager immediately (or in his/her absence a general practitioner or the assistant practice manager) and ensure that an accident reporting form is completed.

In addition the following procedure should be carried out.

1   Immediately following exposure, the site of the exposure (i.e. the wound) should be washed liberally with soap and water but without scrubbing. Exposed mucous membranes including conjunctivae should be irrigated copiously with water after first removing contact lenses if present. If there has been a puncture wound free bleeding should be encouraged gently but the wound should not be sucked.
2   Assess significance of exposure.

Significant exposure includes all percutaneous exposures (skin penetrated or cut by needle or other sharp contaminated with blood or other body fluid) and mucocutaneous exposures to blood (mucocutaneous exposure occurs when the eye(s), insides of nose or mouth or an area of non-intact skin are contaminated by blood).

If exposure is significant:

1   if identifiable, counsel and seek permission from the patient (donor) to take a blood sample for testing for markers of infection (hepatitis B, C, HIV). This will entail pre-test discussion and obtaining fully informed

consent. If the source patient is approached in a sensitive manner consent to testing is rarely withheld. It is important that these procedures are the responsibility of one doctor within the practice who is fully conversant with the *Guidance for Clinical Healthcare Workers: protection against infection with blood-borne viruses*[2] and who has access to urgent advice from a source of expert advice in this area, who may be a consultant in occupational health, virology, microbiology, genito-urinary medicine or public health medicine. The procedures should never be left to the injured healthcare worker (the recipient).

2   check hepatitis B-immune state of recipient from records. Depending on immune status it may be appropriate to offer hepatitis B vaccination or immunoglobulin.[1,2]

3   offer to take a blood sample from the injured healthcare worker for storage and testing in the event of developed illness (5–10 ml clotted sample).

4   if possible the staff member should be offered the opportunity to consult an occupational physician to discuss the implications and risks of developing an illness following the incident. If this is not possible the staff member should be encouraged to consult her or his own general practitioner if she or he has any concerns, and be given details of the action taken after the incident to pass on to the general practitioner.

5   sharps incidents should always be fully documented using a standard in-house recording format and the circumstances investigated in order to identify any failure of protective measure/safety procedures.

6   sharps injuries which lead to illness or a reportable injury should be reported in line with RIDDOR.

Other infections

Tuberculosis

For the protection of staff and patients against tuberculosis (TB) the following are recommended:

1   a pre-employment questionnaire for new staff to assess and possibly exclude those with symptoms which may be due to active infection with TB (e.g. productive cough, weight loss)

2   for staff with a low risk of exposure, no further action

3   for staff at significant risk of exposure through patient contact a Heaf test should be considered if there is no good evidence of past Bacille Calmette-Guèrin (BCG) immunisation (e.g. presence of characteristic scar).[1]

Rubella

At pre-employment stage all healthcare staff should be asked for evidence of immunity to rubella. If there is no evidence of immunity, blood should be taken for rubella antibodies, and if shown to be non-immune, immunisation offered or arranged.

Management of Health and Safety at Work Regulations 1999

These regulations detail how employers should comply with the general duties under HSAWA, and extend the principles of assessing and controlling risks seen in the Control of Substances Hazardous to Health Regulations 1999 (COSHH) into all areas of work activity. Key duties include:

- assessment of all risks to health and safety for employees and anyone else affected by the work activities. This includes risks from chemicals, physical dangers such as manual handling injuries and psychological risks (e.g. from dealing with abusive customers or working long hours)
- recording significant findings
- making arrangements to minimise risk
- appointing competent persons to provide health and safety assistance
- reviewing provision of information and training for employees in the light of the findings of the risk assessment.

When assessing the risks from workplace equipment, it should be noted that electrical equipment is additionally covered by the Electricity at Work Regulations 1995 which state that all electrical equipment must be constructed and maintained to prevent danger so far as is reasonably practicable. Fixed installations such as mains wiring, ceiling lights and wall-mounted heaters should be inspected by a competent electrical contractor who will advise on the necessary re-test interval, which should not exceed 5 years. Other types of equipment may be inspected by a competent person at regular intervals looking at points such as plug and cable condition, fuse rating, general condition of appliances, etc. A system should be in place which ensures that all new equipment is checked before use. Other points to check for would include trailing wires which might cause trips and falls, electrical equipment near to sinks/water and unguarded electrical sockets in waiting areas used by children.

As well as electrical safety it is important to check for non-electrical hazards from surgery equipment, such as ear syringing water-jet machines, audiometers and sterilisers. These should all be maintained in accordance with manufacturers' instructions. Pressure sterilising autoclaves are covered by the Pressure Systems and Transportable Gas Containers Regulations 1989 which require maintenance checks by 'a competent person' at least every 14 months. (The Health and Safety Commission is considering replacing these regulations with the new Pressure Systems Safety Regulations in 2000. If introduced, these will be published with a supporting Approved Code of Practice.) Information about the type of checks required should be available from the manufacturer or supplier.

Display Screen Equipment (DSE) Regulations 1992

Work with visual display unit (VDU) equipment such as VDU screens and other similar equipment is not high risk, but can lead to muscular and other physical problems, eye fatigue and mental stress. The DSE Regulations are aimed primarily at ensuring that these problems are prevented by attention to the correct ergonomic set up of the workstation, the working environment and the tasks performed. In addition there are requirements relating to the provision of eye tests on request and provision of corrective spectacles for VDU use by the employer where these are needed specifically and exclusively for VDU use.

The Regulations apply to all 'regular users' of display screen equipment which could therefore cover most practice employees including the GP partners themselves.

The definition of a VDU 'user' includes anyone using a VDU for continuous periods of an hour or more on most days, who has to transfer information quickly to or from the screen, or anyone using a VDU more or less continuously on most days.

Employers are obliged to assess not only the VDU itself but the whole workstation including seating, environmental factors, the software and task design. Assessments can easily be made by the VDU user him/herself using a standard checklist, with any points identified reviewed in more depth. A suitable checklist is provided in the HSE booklet *VDUs: an easy guide to the Regulations*. Where 'users' request an eye test, employers have to provide an eye test taking into account VDU work, which may be performed by an optician or ophthalmically qualified doctor. The recommended interval is usually 2 years, but employees may ask for a re-test sooner if they have symptoms which they feel may be due to use of the VDU. Corrective spectacles need only be supplied at the employer's expense if needed specifically for correction of vision at the VDU distance.

Manual Handling Regulations 1992

These Regulations apply to any manual handling operations which may cause injury at work. Such operations include not only the lifting of loads but also lowering, pushing, pulling, carrying or moving them, whether by hand or application of other bodily force. The loads involved may be inanimate such as boxes, trolleys, etc., or animate, such as patients. Although the business of general practice is not built around manual handling operations in the way that many workplaces are, there are still a number of tasks which should be assessed under these Regulations.

Fifty-five per cent of reported injuries in the healthcare sector are classified as being caused by manual handling. Compensation payments as high as £203 000 have been awarded to individual nurses whose back injuries were sustained at work. The annual cost to the health service in terms of loss of trained staff, sickness payments and treatment and rehabilitation costs is enormous.

What you need to do

1   Identify all manual handling tasks which may present a risk of injury to the handler. All tasks, including those which are performed only infrequently such as furniture shifting and those which are performed away from the main surgery premises should be considered. Examples might include:
    - lifting portable equipment such as electrocardiograph (ECG) machines/ defibrillators in and out of cars
    - lifting/aiding of patients within their homes
    - lifting stock boxes of office supplies such as photocopier paper.
2   Assess the task, the load, the working environment and individual capability in respect of unavoidable manual handling tasks to identify steps which will eliminate or reduce the risk. Consider the use of hoists or other mechanical lifting aids for frequently performed or high risk operations. The assessment can be most easily done with reference to guidance.[3,4] Following assessment the following actions may need to be taken:
    - prioritise and implement steps needed to reduce risk of injury
    - provide training to all staff which is relevant to their job
    - monitor effectiveness of risk reduction measures and training
    - change attitudes to ensure that employee physical well-being is prioritised in the process of caring for practice patients.

New and expectant mothers

The Management of Health and Safety at Work Regulations 1999 incorporate the risk assessments required by the Pregnant Workers Directive. Regulation 16 requires employers to pay particular attention to potential risks to women who are new or expectant mothers, or who are breast feeding. In a practice setting this may include assessment of the degree of risk of contracting infections which may present risks to the unborn child (e.g. hepatitis B, HIV, herpes, TB, chickenpox, rubella).

Where the risk of contracting an infection in the course of their work is considered to be no greater than the risk of contracting infection by normal (non-work) social interaction, there is no need for any action to be taken. This will generally be the case for non-clinical staff such as secretaries and receptionists.

Other risks which need to be considered and assessed include manual handling activities, particularly in later stages of pregnancy, and involvement in the preparation and administering of cytotoxic drugs, such as for dispensers and clinical staff.

Further information can be found in:

Health and Safety Executive (1994) *New and Expectant Mothers at Work: a guide for employers.* HSE Books, Sudbury.

Health and Safety Executive (1997) *Infections in the Workplace to New and Expectant Mothers.* HSE Books, Sudbury.

Disability Discrimination Act 1995 (DDA)

The section of this Act which relates to employment and places obligations on employers only applies to businesses which employ 15 or more employees. As such it may only apply to larger practices. However, the principles of the Act embody good employment practice and should therefore be followed irrespective of practice size. The Act makes it unlawful for employers to discriminate against current or prospective employees with disabilities on account of their disability. This applies in areas of:

*   recruitment, i.e. at pre-employment assessment
*   promotion
*   training and development
*   dismissal.

The Act covers people with a past disability as well as a present one, and that disability is defined as: 'A physical or mental impairment which has a substantial and long-term adverse effect on a person's ability to carry out normal day-to-day activities'.

Employers must also make reasonable changes to premises, work practices and employment arrangements, e.g. hours of work and work breaks, to accommodate a disabled employee.

Long-term effects include those which have lasted or are likely to last 12 months or more, whether continuously or by periodic recurrence. Normal day-to-day activities cover the following broad categories:

* mobility
* manual dexterity
* physical co-ordination
* continence
* ability to lift, carry or move ordinary objects
* speech, hearing or eyesight
* memory, or ability to concentrate, learn or understand
* being able to recognise physical dangers.

Reporting of Injuries, Diseases and Dangerous Occurrences Regulations 1995 (RIDDOR)

As described in Chapter Three, these regulations require injuries, diseases and occurrences in specific categories to be notified to the enforcing authority by the employer. The duty of notification is on the employer. Immediate notification by telephone should take place where there is a fatal or major injury to anyone sustained in an accident connected with your business. Reporting should otherwise take place by post using standard forms (F2508 for injuries and dangerous occurrences, F2508A for occupational diseases) available from HSE Books for any injury to an employee which results in their absence from work, or being unable to do their normal work for more than 3 days. Also reportable are any of the causes of occupational ill-health suffered by employees listed in the Regulations.

A report should be made to RIDDOR of any general practice staff member who falls at work or injures themselves in other ways (perhaps lifting a patient) and subsequently takes more than 3 days off. Other illnesses which may be relevant in a general practice setting are:

* occupational dermatitis
* occupational asthma

- any illness caused by a pathogen, e.g. hepatitis/TB, reliably attributable to the work activity.

Waste management and duty of care

Under Section 34 of the Environmental Protection Act 1990, a duty of care rests upon those who dispose of waste to make sure that it is safely and legally managed at all stages from production to final disposal. It is necessary to ensure the following.

1   Waste is stored safely and securely while still on practice premises, e.g. sharps awaiting collection stored in suitable sharps bin out of reach of children.
2   The contractor taking away your waste is legally entitled to do so. Office waste will usually be collected by the local council or its contractors. Clinical waste will normally be collected by contractors licensed by the employing health authority/Trust. If using other contractors ensure that they hold appropriate registration as waste carriers and hold a licence to dispose of the types of waste collected.
3   A transfer note is completed and signed by both parties when waste changes hands, and a written description of the waste handed over. Repeated transfers for the same type of work may be covered by one transfer note for periods of up to 1 year. Both parties should keep copies of the transfer note for 2 years. The transfer should include information about:
    - what the waste is
    - what sort of container it is in
    - the time and date of transfer
    - where the transfer took place
    - names and addresses of both parties
    - details of the waste carrier's registration number and the name of the licensing council.
4   Clinical waste and sharps are kept separate from general waste, and in order to comply with the requirements of the Carriage of Dangerous Goods (Classification, Packaging and Labelling) Regulations 1996, should be packed in UN type containers, e.g. identifiable sacks and sharps bins. It is the consigner's duty to ensure that this is done and to provide the contractor collecting the waste with the information required by Regulation 13 of the Carriage of Dangerous Goods (by Road) Regulations 1996. You should discuss this with the contractor so that you know how to comply.

5   Precautions are taken to prevent theft, scavenging or vandalism, for instance outbuildings and store sheds should be locked and free from vermin.

Employment and partnership issues

Pre-employment assessments

All employers are keen to know that prospective employees will be fit to render regular and efficient service, and to ensure that the employee has no medical condition that could place them at greater risk of illness or injury at work. Most of these points can be adequately covered by a well designed pre-employment health questionnaire. If such a questionnaire is introduced the following points should be addressed.

1   What questions should be asked to ensure fitness for the role?

For a practice nurse appointment this may include questions about hepatitis B immune status, TB immunity, spinal fitness; whereas for a data entry clerk questions about previous vision tests and upper limb disorders may be more relevant. Enquiries about medical conditions which have no relevance to the performance of the proposed duties should be avoided on ethical grounds, but questions may be asked which might help you to comply with the Disability Discrimination Act.

2   Who will receive the questionnaire?

The confidentiality of the prospective employee should be respected. One partner within the practice who is aware of the health criteria required is the only person who needs to see the questionnaire. Alternatively, and preferably, an outside agency with knowledge of the health criteria relevant to the work to be undertaken could be asked to review health questionnaires, for example a local occupational health provider or via a reciprocal arrangement with another local practice. This is particularly important in a small close-knit organisation like a general practice where confidentiality is paramount.

3 Is there any need for a medical examination at the pre-employment stage?

In general the answer is no. However, you may wish to perform specific baseline measures of relevance, for instance:

*   vision tests for VDU users/vocational drivers
*   lung function tests for nurses who may be exposed to respiratory sensitisers such as glutaraldehyde
*   review of immunity to infectious agents and immunisation where indicated
*   assessment of specific capabilities where a health questionnaire has highlighted a possible relevant disability.

If it is decided that pre-employment screening of this type is required, it is important that the GPs or their staff do not step outside their role as prospective employers and act as occupational health advisers to the practice due to the ethical conflict that could result. An outside provider of OH services or nearby practice can give an independent viewpoint on fitness.

Sickness absence and ill-health retirement

It is part of good management practice nowadays to actively monitor sickness absence levels and to 'manage' absence. As with pre-employment assessments, it is undesirable for a partner within the practice to act as anything other than the employer in cases of management of sickness absence, rehabilitation after long-term absence or termination of employment on health grounds. In all of these cases an outside provider of OH services should be engaged, with all the necessary information about absence records, nature of duties, hours of work and possible redeployment provided, just as would occur if the employer was not in the business of healthcare. This is just as important whether the ill worker is a receptionist, practice manager or one of the doctors.

Just as it is inadvisable for the employing GP to act as occupational health adviser to his or her general practice team in cases of sickness absence or ill-health in practice staff or partners, it is also inadvisable for the employers to act as the general practitioners for their staff or GP partners because of the dangers of role conflict. It is good practice for all staff and partners to register for general medical services with another local practice where possible.

Stress in general practice

In comparison with 'white collar' and professional workers in industry, health workers of all types report significantly higher levels of pressure at work, lower job satisfaction and less control over their working environments. Doctors in particular have higher than average levels of anxiety and depression, and a greater risk of suicide than the general population.[5,6] Easy access to medications and reluctance to seek help from colleagues may discourage GPs with health problems from seeking help.

While assessing risks to health from physical hazards in the workplace it is also important to assess the risks of stress-related illness to GP partners and practice staff, to acknowledge stress as an issue and to consider measures which will help to reduce the risks. With regard to the partners this may include issues such as:

* encouraging appropriate delegation
* sharing surgery workloads equitably

- considering carefully the taking on of outside practice commitments (extra work = extra income, but at what price?)
- ensuring some protected time each week
- delegating out-of-hours work via on call/weekend working arrangements – use of GP cooperatives, etc.
- use of locums to cover holidays/illness
- sabbaticals and study leave arrangements
- time management and relaxation skills training.

As GPs typically consult their medical advisers only as a last resort, you might consider fostering arrangements whereby each partner attends their own GP on a regular basis for a routine review or medical. Issues such as diet, alcohol, smoking, exercise, coronary risk factors might all be reviewed with an opportunity to 'de-brief' on the past year.

In addition some NHS Trusts are now making their occupational health services available to general practitioners and their staff, some providing a 24-hour counselling service which is useful for post-incident counselling where partners or staff have dealt with distressing events such as major incidents and acts of violence.

The British Medical Association set up a telephone stress counselling service for BMA members in 1996 (tel: 0645 200169). The service received more than 3000 calls in its first year of operation, the most commonly raised issues being emotional health, anxiety and workload.

The National Counselling Service for Sick Doctors was set up in 1985, enabling doctors to refer themselves or seek advice about colleagues in a confidential way (tel: 0870 2410 535) from Monday to Friday between 9.30 a.m. and 16.30 p.m. On contacting the service the enquirer is given the name of a medically qualified adviser who will offer advice or arrange for a medically qualified counsellor to help.

Members of the practice staff other than the GPs may suffer from work-related stress. Measures which may help to minimise the risks include the following:

- assessing work systems and activities to identify whether staff could be exposed to possible risks of work-related stress
- ensuring tasks are achievable
- managing staff consistently and fairly
- communicating effectively
- managing change effectively
- a culture of open acknowledgement of stress as a risk for practice staff
- training in relevant issues, e.g. stress management, relaxation techniques
- team building and social events for staff
- provision of access to outside occupational health service or an Employee Assistance Programme
- opportunities for staff to contribute to practice development.[7]

Violence to staff

Violence in the context of the healthcare sector is defined as follows: 'Any incident in which a person working in the healthcare sector is abused, threatened or assaulted in circumstances relating to their work'.

Using this broad definition violent incidents do not need to cause physical harm and would include, for example, a receptionist who was verbally abused by an irate patient annoyed at waiting for 'too long' to be seen. Incidents which occur outside the surgery premises are also included, such as when GP partners and nurses are on home visits, office staff taking letters to the post office, etc.

Factors which may increase the risk of violence include:

- working alone
- working after normal hours
- working and travelling in the community
- carrying valuable equipment or medicines
- visiting patients who are emotionally or mentally unstable
- patients under the influence of drink or drugs.

The British Crime Survey (1995)[8] confirmed that health professionals appear to be at higher risk of work-related violence than the general population.

Occupation	Incidents per 10 000
Medical practitioners	762
Nurses and midwives	580
Other health-related occupations	830
All survey subjects	251

The framework of risk assessment and management previously outlined at the beginning of this chapter can be applied to the risk of violence to staff in exactly the same way as to more conventional risks. The assessment procedure may identify groups of patients who present a greater risk, activities which present a greater risk and staff who are at greater risk.

Measures which may be considered to reduce or eliminate risk of violence include:

- premises: location and number of entrances
- lighting
- comfort and space in waiting areas
- fixtures and fittings (fixed or heavy chairs in waiting areas which cannot be used as weapons)
- provision of information such as likely waiting times

- facilities such as public telephones or play areas, etc.
- alarm or panic buttons in consulting and treatment rooms
- working patterns and practices
- staffing levels
- training: this is appropriate for all staff at risk from violence and should typically cover:
  - causes of violence
  - recognition of warning signs
  - relevant interpersonal skills
  - details of working practices and control measures
  - incident reporting procedures
- personal security:
  - personal alarms
  - plans of any visits and movements
  - periodic reporting in
  - generic assessments of areas and patient groups
  - mobile phones
  - consideration of need for joint visits.

Reporting/debriefing after incidents may help to prevent future incidents and relieve stress.

Remember that there is a requirement to report incidents of violence to employees under the RIDDOR Regulations where the incident results in death, major injury, or incapacity for normal work for 3 days or more.

Health promotion for GPs and their staff

GPs should make all efforts to maintain and improve the health of their staff and themselves by taking part in national and local health promotion initiatives, by encouraging healthy lifestyles (e.g. discounted group membership of a local gym) and by making use of their easy access to health promotion resources such as talks for staff from physiotherapists on 'back care' or from practice counsellors on 'stress reduction techniques'.

Role of Health and Safety Executive and the Employment Medical Advisory Service[9]

The Health and Safety Executive is the relevant enforcing agency established by the Health and Safety at Work, etc. Act. Its medical branch, the Medical

Inspectorate, is staffed by occupational physicians and nurses. As well as helping the enforcement agency where medical issues are involved, they can provide advice to employers, employees and the general public.

The general approach of HSE inspectors is to give advice if deficiencies are found during an inspection of a workplace, with follow up to ensure action is taken. In more serious cases, improvement notices may be served requiring the employer to take specific action within a specified time. If they feel there is immediate danger a prohibition notice may be served, legally preventing the employer from undertaking the dangerous activity.

Prosecution under the Health and Safety at Work, etc. Act is also possible, either of the organisation or the individual charged with responsibility for health and safety. Successful prosecution carries with it a criminal record and an automatic report to the General Medical Council if doctors are involved. If the case is dealt with in a Magistrates court the penalty may be a fine or referral to the Crown Court. The Crown Court may impose an unlimited fine and in some cases a custodial sentence of up to 2 years.

References

1   Department of Health (1996) *Immunisation Against Infectious Disease 1996.* HMSO, London.

2   Department of Health (1998) *Guidance for Clinical Healthcare Workers: protection against infection with blood-borne viruses. Recommendations of the Expert Advisory Group.* Department of Health, Wetherby (available on www.open.gov.uk/doh/chcguid1.htm).

3   Health and Safety Executive (1992) *Manual Handling Operations Regulations: guidance on regulations.* HSE Books, Sudbury.

4   Health and Safety Executive (1998) *Manual Handling of Loads in the Health Services.* HSE Books, Sudbury.

5   Chambers R and Campbell I (1996) Anxiety and depression in general practitioners: associations with type of practice, fundholding, gender and other personal characteristics. *Fam Pract.* **13**(2): 170–3.

6   Lindeman S, Laara E, Hakko H and Lonnqvist J (1996) A systematic review on gender-specific suicide mortality in medical doctors. *Br J Psych.* **168**: 274–9.

7   Health and Safety Executive (1995) *Stress at Work: a guide for employers.* HSG 116. HSE Books, Sudbury.

8   Home Office Research and Statistics Department (1995) *British Crime Survey.* HMSO, London.

9   Health and Safety Executive (2000) *The Employment Medical Advisory Service and You.* HSE5. HSE Books, Sudbury.

Reflection exercises

If you want to consolidate your learning, you need to spend some time thinking about how what you have read applies to your own practice. Why not do one, or all, of the following exercises?

Read through the extracts of the report and recommendations of the Occupational Health Advisory Committee on *Improving Access to Occupational Health Support* before you start – to extend your knowledge and thinking about the quality of your occupational health services and the extent to which you meet your workers' needs (*see* Appendix 2).

Exercise 1. Assess the content and application of your health and safety policy. First, find your practice's health and safety policy.

(a)   How easy was it to find your policy?
(b)   When was it last reviewed/amended?
(c)   Who performed the last modification?
(d)   Compare your policy with the example outlined in this chapter; how does it differ?
(e)   Look at the extent to which your practice is complying with health and safety legislation – compare the application of your policy against the checklist below.

Practice compliance with current Health and Safety Legislation*

Agreed standard	*Standard achieved*
1 The practice has a written health and safety policy which complies with current legislation. The policy includes:	Yes/No
(i) written statement of employer's commitment to health and safety, signed by GP and dated	Yes/No
(ii) responsibility for implementing and monitoring policy defined for each aspect of the policy	Yes/No
(iii) procedures and responsibilities specified for identifying and controlling particular hazards in the workplace (e.g. control of drugs, medicines, syringes, etc.).	Yes/No
2 The policy is reviewed annually or more frequently if changes in work practices or legislation.	Yes/No
3 The health and safety policy includes reference to:	
(i) provision and maintenance of equipment that is safe and without risks to health	Yes/No

continued opposite

(ii)	the safe use, handling, storage and transport of inherently or potentially dangerous articles and substances	Yes/No
(iii)	every member of the practice receives adequate information, instruction, training and supervision on health and safety matters	Yes/No
(iv)	safe means of access to and exit from the practice	Yes/No
(v)	the practice display the employer's liability insurance certificate in a public place.	Yes/No

---

4	Each member of the practice team is aware of the requirements of the practice health and safety policy.	Yes/No

---

5	The practice carries out a systematic risk assessment of workplace risks by the 'competent person' who:	Yes/No
	• makes arrangements to put into practice preventive and protective measures – planning, monitoring, organisation, training, control and review	
	• must prioritise any measures required	
	• sets up procedures to deal with dangers and hazards.	

6	The practice records all incidents and notifies serious accidents or occupational-related diseases to the Health and Safety Executive (in line with Reporting of Injuries, Diseases and Dangerous Occurrences Regulations 1985, RIDDOR).	Yes/No

---

7	The practice displays notices warning of any particular hazards.	Yes/No

---

8	The practice checks that equipment and facilities conform to current health and safety requirements.	Yes/No

---

9	The practice has a first aid box which is well stocked, readily accessible, and includes eye wash equipment.	Yes/No

---

10	Preventive measures taken where doctors and staff perceive that they are at risk from violent behaviour; appropriate training arranged.	Yes/No

---

11	The practice's safety policy addresses fire safety, fire prevention and fire procedures, including:	Yes/No
(a)	Fire certificate or written evidence of approval of local authority fire officer	Yes/No
(b)	nominated fire officer and, if appropriate, people responsible for specific actions in the event of a fire	Yes/No
(c)	regular fire training and evacuation drills for staff, at least annually	Yes/No
(d)	ensuring that fire exits are unobstructed, clearly marked and inspected daily	Yes/No
(e)	regular testing of fire warning systems based on manufacturer's instructions	Yes/No
(f)	sufficient and appropriate fire-fighting equipment available and maintained on a maintenance contract.	Yes/No

---

* Framework for this checklist and standards is a modified version of: Calderdale and Kirklees Health Authority (1999) *Quality Standards Framework for Primary Health Care Teams*. Calderdale and Kirklees Health Authority, Leeds.

What changes will you make now? And when will you check that the changes you intend are in place?

This Exercise might take 1 hour.

Exercise 2. Use of visual display units (VDUs)

(a)   How many VDU screens are there in your practice?
(b)   And how many VDU users are there?
(c)   What are the adverse health effects caused by work with VDUs?

This Exercise might take 1 hour.

Exercise 3. The potential threat of violence and aggression to staff
Can you think of some examples for each of the following?

(a)   Five patient groups who may present greater risks of violence.
(b)   Six workplace activities in or about the surgery which might expose
       workers to greater risks of violence.
(c)   Four categories of medical support staff at greater risk of
       violence.
(d)   Describe four practical measures which can be developed to prevent and
       reduce violence in and about the practice.

This Exercise might take 1 hour.

Exercise 4. Risk reduction of hazards of lifting
How might you monitor the effectiveness of your training and risk reduction measures with regards to manual handling?

This Exercise might take 1 hour.

Exercise 5. Health promotion
Write down five ideas for health promotion activities that are relevant to those who work in your practice.

   Discuss how you might develop one of these ideas into an occupationally linked community 'partnership' initiative; that is, linking the activities and/or workforce in your practice with those in another non-health organisation.

This Exercise might take 1 hour.

CHAPTER EIGHT

Designing your personal development plan, using stress management as an example

Your personal development plan[1,2]

Your personal development plan will allow you to demonstrate your fitness to practise or manage one or more aspects of occupational health, whether you are a GP, nurse or practice manager. It might form the major part of a future revalidation programme for GPs.

Your plan should:

- identify your weaknesses in knowledge, skills or attitudes
- specify topics for learning as a result of changes: in your role, responsibilities, the organisation
- link into the learning needs of others in your workplace or team of colleagues
- tie in with the service development priorities of your practice, the Primary Care Group/Trust (PCG/T), your district or the NHS as a whole
- describe how you identified your learning needs
- prioritise and set your learning needs and associated goals
- justify your selection of learning goals
- describe how you will achieve your goals and over what time period
- describe how you will evaluate learning outcomes.

You will need to set aside enough time to shape and justify your learning plan. The more time you invest in making your plan and the programme of learning, the more likely it is that you will focus your learning effectively.

The main task is to capture what you have learned in some way that suits you, so that you can look back at what you have done and:

- reflect on it at a later date, to learn more, make changes as a result, identify further needs
- demonstrate to others that you are fit to practise or work: through what you have done, what you have learnt and what changes you have made as a result; the standards of work you have achieved and are maintaining; how you monitor your performance at work
- use it to show how your personal learning fits in with the practice business, and the practice personal and professional development plans.

Use a range of methods to identify your learning needs[2]

Your learning needs will encompass the context in which you work as well as your knowledge and skills in relation to any particular role or responsibility of your current post. Your learning needs in relation to stress management will be different depending on whether you practise or work in an inner city compared to a rural location.

Use several methods to identify your learning needs. No one method will give you reliable information about the gaps in your knowledge, skills or attitudes.

1 *Appraise yourself – review how you work.* This will be a mix of skills, knowledge and personal attributes that you need; or new processes you anticipate coming into being in your workplace.

Your, the practice's, the PCG/T's aspirations for:

- new models of service delivery
- new roles or responsibilities in the organisation
- organisation's vision for change.

Your attitudes to:

- other disciplines
- patients
- life-long learning
- culture
- change.

Context of work

- networking in health and non-health settings
- team relationships
- different sub-groups of the population
- historical service provision
- the organisation's priorities.

Your knowledge

- clinical
- about your local population
- about your organisation
- local experts or other provision
- of best practice
- range of services available locally
- systems and procedures in your organisation
- inequalities of health or healthcare of your patient population.

Legal requirements

- health and safety at work
- new legislation
- employment procedures, e.g. equal opportunities
- safe practices, e.g. personal safety.

Awareness of health policies

- new health policies
- national priorities
- local priorities, e.g. health improvement programme
- fashions, e.g. in clinical practice or how education is delivered.

Skills

- team working and communication
- effective working practices
- your basic competence
- health needs assessment
- practice management
- organisational development of your team
- information technology and computer capability
- your specialist areas
- planning
- personal management.

   2 *Ask other people what they think of you: gain feedback from colleagues.* Feedback from colleagues or mentors can help you to see what it is that you do not know that you don't know.

▼
Asking colleagues what they think of you can be illuminating.

3 *Audit methods*

- Peer review.
- Criteria based audit.
- External audit.
- Direct observation.
- Surveys.
- Significant event audit – of a stressful event for you that happened recently.

4 *Monitor your own or your practice's clinical decisions.* For example, review the extent to which you adhere to pre-agreed clinical protocols, guidelines and care pathways. Does being unable to justify deviations from the agreed procedures reveal any learning needs – are you stressed by there being so much to learn or remember?

5 *Monitoring the process of obtaining healthcare – that is, access, availability, satisfaction.* Will audit identify causes or effects of stress?

- Access and availability.
- Patient satisfaction.
- Referrals to other agencies and hospitals.
- Patient complaint.

6 *Monitoring systems and procedures.*

7 *Informal conversations – in the corridor, over coffee.*

8 *Strengths, weaknesses, opportunities and threats (SWOT) analysis.* Brainstorm the strengths, weaknesses, opportunities and threats of your situation in the practice, or the practice as a whole. Strengths and weaknesses of individual practitioners might include: knowledge, experience, expertise, decision making, communication skills, inter-professional relationships, political skills, timekeeping, organisational skills, teaching skills, research skills. Strengths and weaknesses of the practice organisation might relate to most of these aspects too as well as resources – staff, skills, structural.

Opportunities might relate to unexploited potential strengths, expected changes, options for career development pathways, hobbies and interests that might usefully be expanded. Threats will include factors and circumstances that prevent you from achieving your aims for personal, professional and practice development.

Prioritise important factors. Draw up goals and a timed action plan.

9 *Compare your performance with externally set standards.*

10 *Observation of your work environment and role.* This could be informal and opportunistic, or more systematic, working through a structured checklist. Do you know what other colleagues do? How do their roles and responsibilities interface with yours? Ask others what they think of your performance.

11 *Reading and reflecting.*

12 *Look at your practice population's health needs,* and see what you may need to learn as an individual or practice team to address those needs more effectively.

13 *Review the business or development plan of your practice* or primary care group/trust and other official strategic documents or directives.

14 *Job appraisal.*

15 *Educational appraisal.*

16 *Review practice protocols and guidelines.*

Integrating your personal development plan with the practice personal and professional development plan[2]

The practice *personal and professional development plan* (PPDP) should cater for everyone who works in a practice. Clinical governance principles will balance the development needs of the population, the practice, the PCG/T *and* your individual *personal development plan* (PDP).

You might want to start by identifying your own learning needs, combining them with other people's and then checking them against the practice business plan. You could start from the other direction – develop a practice-based PPDP from your business plan and then identify your individual learning needs within that. Whichever direction you start from, you must ensure that you integrate your individual needs with those of your practice, the wider community and the needs and directives of the NHS.

Your learning plan should complement the professional development of other individuals and the development plan for the practice. If you are working on a project that involves change for other people as well as yourself, it is better to work together towards a common goal and coordinate multi-professional learning across traditional boundaries.

If you work in a number of different roles or posts, gaps and duplication of activities should be avoided. After reflection about the boundaries between your roles you may be able to focus your learning so that meeting your needs in one role benefits another.

Make your learning plan flexible; you may want to add something in later when circumstances suddenly change or an additional need becomes apparent – perhaps as the result of complaint or hearing something new at a meeting.

Long-term locums (say, for longer than 6 months), assistants, retained doctors and salaried GPs should all be included in the practice plan. Remember to include all those staff who work for the practice however few their hours – you cannot manage without them or they would not be there!

Time is one of the resources that must be considered when drawing up your action plan – be realistic or you will abandon your plan halfway through with little achieved. Adequate resources must be in place for your learning needs and protected time must be built in. If you need additional funds for your practice learning plan, that is usually best done as part of the practice development plan by an identified person who can contact the clinical governance or education lead of the PCG/T about possible sources of help.

Your personal development plan
will feed into a
practice personal and professional development plan
that will feed into the
education and clinical governance programmes for your primary care group or primary care trust.

Worked example of a personal development plan, focused on stress management

The worked example demonstrates how you might set about preparing and undertaking your own programme based on stress management. Alternatively, you might have chosen a more clinically based topic such as one or two commonly occurring occupational diseases or the whole field of occupational medicine in relation to the effects of work on your patients' health.

A personal development plan should contain:[2]

- a priority topic justified by a previous needs assessment
- an action plan
- measure of baseline and follow-up level of knowledge, skills, attitudes, etc. to allow evaluation of learning
- methods of learning relevant to topic
- how learning is incorporated into everyday practice and disseminated to others
- further learning required.

Modify the worked example around stress management to match your own ideas and circumstances. You might find that a personal development plan around one topic is all that you can manage in the course of one year, especially if you widen your programme around the various components of clinical governance that you incorporate into your plan. For instance, considering the evidence base in relation to 'stress management' might lead you search for and apply the evidence on various psychological and alternative therapies for patients.

The worked example is followed by a template for you to prepare your own plan – photocopy the pages – and get on with it! Demonstrate what you have achieved and keep a learning record.

Worked example of personal development plan: health at work

What topic: stress management

Justify why topic is a priority

A personal and professional priority? After working as a GP for 20 years I realise that my job satisfaction is being whittled away by the stresses I am experiencing at work from the many demands on me. I think that the effects of this stress are starting to threaten the quality of my everyday work.

A practice priority? A GP being under stress affects all members of the practice team. This may be because the stressed GP creates more work for others if their performance is under par such as being forgetful, or making mistakes; or stress may mean resistance to change.

A district priority? The PCG is encouraging more awareness of the need for, and benefits of, good stress management in general practice.

A national priority? The government is emphasising the importance of good human resource management in the quality and well-being of the NHS workforce.

Who will be included in my personal plan?

Although my personal development plan is focused on my needs, I cannot 'beat' stress without significant changes in the organisation of the practice. So I will invite others to join in my initiative as it evolves, either learning to control stress for themselves or involving them in reorganising the practice systems that create stress. Those included will be:

- other GPs
- practice manager
- practice nurses
- reception staff
- practice secretary
- family and partner at home.

What baseline information will I collect?

Causes and effects of stress on me at work, and outside work.

How will I identify my learning needs?

1   Audit of my everyday practice, e.g. significant event audits of several unexpected demands – extra patients, interruptions – whatever crops up over a couple of days.
2   Observation of my practice: by me using a stress log diary and self-assessment scores of perceived stress; and by informal comments from others.
3   Observation of stressors in my life outside work: by me using a stress log diary; and by informal comments from others.

continued opposite

4   Comparing what methods of stress management I know about against the list of possibilities given in a manual on stress management (e.g. *Survival Skills for GPs* and *Survival Skills for Nurses*[3,4]).

5   Group discussion in practice meeting attended by GPs, employed and attached practice staff where the general topic of 'stress at work' is debated. Should learn more about causes of stress for me by listening to what others find stressful and hearing more about their concerns and feelings.

What are the learning needs for the practice and how do they match my needs?
The practice will benefit if I am less stressed as an individual GP when I am more likely to take a lead or actively support changes to improve practice systems and procedures. If I learn to control some of the stresses upon me, this should have additional benefits for others in the practice team as I should be easier to work with, more efficient and a better communicator.

I shall have to take care that my suggestions for improving the practice organisation reduce stress levels for everyone whenever possible; if I simply control stress upon me by redirecting demands on others who do not have the time, training or inclination for absorbing those demands, then I would be implementing my personal development plan at the expense of the overall good of the practice, which is untenable.

Is there any patient or public input to my plan?
I will use any informal feedback from patients, or any formal patient complaint about the practice if it is relevant to me; or that information might be the basis for a significant event audit. Unsolicited patient feedback might identify pressures in the practice previously unknown to me, or make me think about the causes or effects of stress for me, from which I might learn, e.g. remarks about punctuality, problems obtaining help or advice.

Aims of my personal development plan arising from the preliminary data-gathering exercise
To reduce stress at work by:

• identifying three significant sources of stress for me at work that are within my ability to control as an individual or working with others in the practice
• learning how to recognise causes of stress and their effects on me
• learning more about methods of stress management appropriate for the stressors I have identified and how to apply them
• learning how improving practice systems and procedures might be possible and how to involve others in the practice team in such reorganisation.

continued overleaf

How I might integrate the 14 components of clinical governance into my personal development plan focusing on stress management[5]

Establishing a learning culture: meet with another GP who is doing her/his own personal development plan to swap notes and encourage each other.

Managing resources and services: control key sources of stress such as reducing the impact of patients needing to be seen on the same day as 'urgent' cases or reducing interruptions.

Establishing a research and development culture: carry out a survey to identify sources of stress, or compare levels of demands before and after introducing an intervention.

Reliable and accurate data: becoming more competent at operating the practice computer may be a possible solution to reducing stress, revealed by the preliminary needs assessment.

Evidence-based practice and policy: updating my knowledge of the evidence for best practice for common clinical conditions may relieve feelings of guilt and uncertainty causing me stress, or reduce the potential for making mistakes.

Confidentiality: re-register as a patient with a nearby practice where my medical details will be confidential; some colleagues have access to my records in the practice where I work.

Health gain: reducing stress will result in less physical symptoms of stress and associated medication such as analgesics, indigestion remedies, etc.

Coherent team: good communication between me and the others in the practice team should reduce stress for us all – perhaps through a news-sheet when changes are expected.

Audit and evaluation: audit will be part of the learning needs assessment as described; the beneficial (hopefully) effects of any intervention will be evaluated.

Meaningful involvement of patients and the public: seeking patients' views through focus groups (for instance) might help us improve access so that I find better ways of booking patients so that I am not under such time pressure when they consult me.

Health promotion: as I learn what works for me in managing stress I might ensure that similar help is available to promote stress reduction for my patients, e.g. relaxation tapes.

continued opposite

Risk management: using a risk assessment, reduction and management approach to my personal safety in the practice and on home visits should identify, avoid or minimise risk factors which threaten my (and others) safety and provoke stress.

Accountability and performance: being under too much stress for too long a time, will inevitably make me less effective; this should be reversible with good stress management.

Core requirements: I am more likely to adopt new approaches that are more cost-effective if I am less stressed and more willing to embrace change.

Action plan

Who is involved/setting: me in the general practice setting plus anyone else in the practice team or associated with it, with whom I work.

Timetabled action: start vv date

by xx month: preliminary data gathering completed and any others involved in initiative
- is there a practice protocol for managing stress?
- map expertise in practice (e.g. community psychiatric nurse, practice nurse with counselling skills); list other providers of help (e.g. BMA stress counselling service)
- baseline information about sources of stress from completed stress.

by xy month: review current performance:
- audits of actual performance via pre-agreed criteria, e.g. numbers of interruptions whilst consulting in surgery; numbers of 'extra' patients seen in addition to booked surgeries
- compare performance with any or several of the 14 components of clinical governance described on previous page.

by yy month: identify solutions and associated training needs:
- learn new skills – in assertiveness, time management, delegation
- write or revise the practice protocol on stress management – to include health surveillance, monitoring sources of stress at work
- clarify my role in the practice organisation – be more definite so I know what my responsibilities are and won't feel guilty when others do not fulfil their duties
- apply the practice protocol for stress management, identify gaps, propose changes to others at practice meeting
- attend external courses or in-house training as appropriate
- visit another practice to see how they have combated their stressors.

continued overleaf

by yz month: make changes:
- feedback information to practice team about what is needed to reduce stress for me – and the others prior to making changes, e.g. set up more opportunities for mutual support
- improve efficiency of practice organisation – from patients' and staff perspectives
- news-sheet once every quarter describing any changes in practice protocol or peoples' roles and responsibilities
- training for practice nurse to whom I am delegating more chronic disease management
- find 'buddy' outside practice with whom to discuss progress on our personal development plans
- re-register with nearby practice and GP in whom I have complete trust and who is independent of my own practice.

Expected outcomes: more effective control of sources of stress; more efficient practice organisation; better teamwork including communication and delegation; more effective performance at work; more willing to consider and implement change.

How does my learning plan tie in with other strategic plans?
It will tie in with the practice's development plan and the PCG/T's development plan as far as possible. The PCG/T plans should in turn match with the local HImP, Social Services and other NHS Trusts' strategic developments. All these organisations encourage ways to be found to manage stress at work better.

What additional resources will I require to execute your plan and from where do I hope to obtain them?
I will ask the practice to sponsor the costs of obtaining relaxation tapes for patients (which I will use too). I would expect to be able to fit attending a stress management course into my working day; and to pay any course fee myself, claiming postgraduate education allowance to cover my costs.

How will I evaluate my learning plan?
I will use similar methods to those I used to identify my learning needs as given earlier: keep other stress logs of the pressures I perceive at work and outside work; re-audit interruptions and the booking times after we have made changes and improved the practice organisation. I will discuss with my 'buddy' how he or she thinks I am progressing with my personal development plan.

How will I know when I have achieved my objectives?
I will re-audit and monitor my stress levels through a record log as described, 12 months later. I will determine whether I am using more effective coping methods to minimise any sources of stress that I have not been able to obliterate.

continued opposite

How will I disseminate the learning from my plan to the rest of the practice team and patients? How will I sustain my new found knowledge or skills?
I will share what I have learnt at the peer support meeting I have set up in the practice for other team members. I will encourage the rest of the practice team to adopt 'health at work' for a practice personal and professional development plan, or at least for it to be the topic of an educational meeting for all the practice team. I will write an article for the medical press on effective stress management in primary care!

How will I handle new learning requirements as they crop up?
I should jot down any thoughts about what else I need to learn about managing stress as I discover it – or I will not be able to remember what it was I was going to look up or think more about, later on.

Check list to be sure that the topic chosen is a priority and the way in which I plan to learn about it is appropriate

Your topic: Stress management

How have you identified your learning need(s)?

a PCG requirement ❏	*e* Appraisal need ❏
b Practice business plan ❏	*f* New to post ❏
c Legal mandatory requirement ❏	*g* Individual decision ☒
d Job requirement ❏	*h* Patient feedback ☒
	i Other ☒

Have you discussed or planned your learning needs with anyone else?

Yes ☒   No ❏   If so, who? *My 'buddy' with whom I am discussing progress with my personal development plan; the practice manager.*

What are the learning need(s) and/or objective(s) in terms of:

Knowledge: What new information do you hope to gain to help you do this? *Sources of stress; effective methods of stress management.*

Skills: What should you be able to do differently as a result of undertaking this development? *Be more assertive; learn to delegate without putting strain on others; introduce change to practice systems so that the organisation is more efficient.*

Behaviour/professional practice – how will this impact on the way you then do things? *My performance at work should improve.*

Details and date of desired development activity. *Attend stress management course that covers a range of coping methods next week.*

Details of any previous training and/or experience you have in this area/dates. *Read widely on the topic; discussed the topic with the community psychiatric nurse over a cup of coffee.*

Your current performance in this area against the requirements of your job:

Need significant development in this area	☒	Need some development in this area	❏
Satisfactory in this area	❏	Do well in this area	❏

continued opposite

Level of job relevance this area has to your role and responsibilities:

Has no relevance to job	❐	Has some relevance	❐
Relevant to job	❐	Very relevant	☒
Essential to job	❐		

Describe what aspect of your job and how the proposed education/training is relevant. *Reducing stress should help me to improve my concentration and perform-ance in almost every aspect of my job as a GP; it should help my home life too if I am less distracted by stress and jobs I have brought home as I seem so inefficient at work.*

Additional support in identifying a suitable development activity?
Yes ❐   No ☒

Describe the differences or improvements for you, your practice, PCG/T as a result of undertaking this activity? *Everyone in the practice and patients should benefit from the new improved version of me when under less stress; they should also benefit if I introduce changes in the practice systems that reduce stress for others, and not just for me. The patients and the PCG may notice that the practice is working more effectively.*

Determine the priority of your proposed educational/training activity:
Urgent ❐   High ☒   Medium ❐   Low ❐

Describe how the proposed activity will meet your learning needs rather than any other type of course or training on the topic. *It is local so I won't waste time in travelling. It is interactive small group work, so I should benefit from hearing about others' ways of controlling stress at work.*

If you had a free choice would you want to learn this? *Yes/No*
    If **not**, why not? (please circle all that apply):

- waste of time
- already done it
- not relevant to my work, career goals
- other:

If **yes**, what reasons are most important to you (put in rank order):

- improve my performance   1
- increase my knowledge   4
- get promotion
- just interested
- be better than my colleagues
- do a more interesting job
- be more confident   2
- it will help me   3
- other:

Record of my learning: about stress management (mid-way through completion of personal development plan)

	Activity 1: dealing with stress at work	Activity 2: time management	Activity 3: increasing support to beat stress	Activity 4: identifying stressors and sources of pressure
In-house formal learning	Community psychiatric nurse ran 1-hour session on dealing with post-traumatic stress, after member of staff died in a road accident		Organised first 1-hour meeting at lunchtime to discuss setting up regular support forum – facilitated by local lay counsellor	
External courses	One day course at nearby postgraduate centre, small group learning – included several stress management methods	Learnt about time management at same 1-day course (Activity 1)	Learnt about support from 1-day course (Activity 1)	
Informal and personal	Chat with practice staff over coffee, in corridors, etc	Spent 2 hours peer assessing other practice as part of GP training activities. Gained new ideas on time management from their practice manager	Read up on the topic from good manual on topic.[3] Did some of the interactive exercises	Fed audit results back to other GPs, practice manager and staff at practice meeting. Led discussion on what we will do about the problem areas – to submit action plan. Interesting discussion afterwards
Qualifications and/or experience gained?	PGEA for 1-day course	Experience gained from comparing my practice with the other training practice		

References

1 Field S (2000) *Continuing Professional Development in General Practice: the regional strategy*. West Midlands Region. West Midlands Postgraduate GP Education Unit, Birmingham.

2 Wakley G, Chambers R and Field S (2000) *Continuing Professional Development in Primary Care: making it happen*. Radcliffe Medical Press, Oxford.

3 Chambers R (1999) *Survival Skills for GPs*. Radcliffe Medical Press, Oxford.

4 Chambers R, Hawksley B and Ramgopal T (1999) *Survival Skills for Nurses*. Radcliffe Medical Press, Oxford.

5 Chambers R and Wakley G (2000) *Making Clinical Governance Work for You*. Radcliffe Medical Press, Oxford.

Reflection and planning exercise

Now complete your own personal development plan. It might be focused on something other than 'stress management', related to other topics in occupational health that have been included in this book. It might be your personal perspective of some of the lines of activity described in the worked example of the practice personal and professional development plan described in the next chapter.

Photocopy the following template or complete the version in the book. Choose a topic that meets your individual needs, or duplicate the plan for stress management. You might choose to look at occupational healthcare of your patients as a wide-ranging topic; or you might focus down on one condition or an aspect such as risk management, or stress management as given.

Choose several methods to justify the topic you have chosen or to identify your learning needs. Incorporate learning needs or baseline information from 'Reflection tasks' at the end of each chapter such as the audit of patient records from the material in Chapter One or the SWOT analysis in Chapter Four. Draft the action plan – you might already have prepared this in the reflection task at the end of Chapter Four. Show it to someone else and get their views as to whether it is relevant and well-balanced and achievable. Plan and undertake your learning and demonstrate the subsequent improvements in your knowledge and practice.

This Exercise might take 10 to 40 hours depending on what topic you choose and the extent of preliminary needs assessment and evaluation that you do.

Template for your personal professional development plan: photocopy the pages and complete one chart per topic

What topic:

Justify why topic is a priority:

A personal and professional priority?

A practice priority?

A district priority?

A national priority?

Who will be included in your personal plan? (anyone other than you? other GPs, employed staff, attached staff, others from outside the practice, patients?)

What baseline information will you collect and how?

How will you identify your learning needs? How will you obtain this and who will do it: self-completion checklists, discussion, appraisal, audit, patient feedback? Look back to the section on identifying your learning needs in the first section of this chapter.

continued opposite

What are the learning needs for the practice and how do they match your needs?

Is there any patient or public input to your personal development plan?

How might you integrate the 14 components of clinical governance into your personal development plan focusing on the topic of XX?

Establishing a learning culture:

Managing resources and services:

Establishing a research and development culture:

Reliable and accurate data:

Evidence-based practice and policy:

Confidentiality:

Health gain:

Coherent team:

Audit and evaluation:

continued overleaf

Meaningful involvement of patients and the public:

Health promotion:

Risk management:

Accountability and performance:

Core requirements:

Objectives of your personal development plan arising from the preliminary data-gathering exercise

Action plan (include objectives, timetabled action, expected outcomes)

continued opposite

How does your personal development plan tie in with your other strategic plans (for example the practice's business or development plan, the Primary Care Investment Plan)?

What additional resources will you require to execute your plan and from where do you hope to obtain them? (Will you have to pay any course fees; will you be able to organise any protected time for learning in working hours?)

How will you evaluate your personal development plan?

How will you know when you have achieved your objectives? (How will you measure success?)

How will you disseminate the learning from your plan to the rest of the practice team and patients? How will you sustain your new-found knowledge or skills?

How will you handle new learning requirements as they crop up?

Check out whether the topic you choose to learn is a priority and the way in which you plan to learn about it is appropriate. Photocopy this proforma for future use

Your topic:

How have you identified your learning need(s)?

a PCG requirement ❐	*e* Appraisal need ❐
b Practice business plan ❐	*f* New to post ❐
c Legal mandatory requirement ❐	*g* Individual decision ❐
d Job requirement ❐	*h* Patient feedback ❐
	i Other ❐

Have you discussed or planned your learning needs with anyone else?

Yes ❐   No ❐   If so, who?

What are the learning need(s) and/or objective(s) in terms of:

Knowledge: What new information do you hope to gain to help you do this?

Skills: What should you be able to do differently as a result of undertaking this development?

Behaviour/professional practice – how will this impact on the way you then do things?

Details and date of desired development activity.

Details of any previous training and/or experience you have in this area with dates.

Your current performance in this area against the requirements of your job:

Need significant development in this area ❐	Need some development in this area ❐
Satisfactory in this area ❐	Do well in this area ❐

Level of job relevance this area has to your role and responsibilities:

Has no relevance to job ❐	Has some relevance ❐
Relevant to job ❐	Very relevant ❐
Essential to job ❐	

continued opposite

Describe what aspect of your job and how the proposed education/training is relevant.

Additional support in identifying a suitable development activity?
Yes ☐   No ☐

If yes, what do you need?

Describe the differences or improvements for you, your practice, PCG/T as a result of undertaking this activity?

Determine the priority of your proposed educational/training activity:
Urgent ☐   High ☐   Medium ☐   Low ☐

Describe how the proposed activity will meet your learning needs rather than any other type of course or training on the topic.

If you had a free choice would you want to learn this? *Yes/No*
   If **not**, why not? (please circle all that apply):

- waste of time
- already done it
- not relevant to my work, career goals
- other:

If **yes**, what reasons are most important to you (put in rank order):

- improve my performance
- increase my knowledge
- get promotion
- just interested
- be better than my colleagues
- do a more interesting job
- be more confident
- it will help me
- other:

Record of your learning: write in topic, date, time spent, type of learning

	Activity 1	Activity 2	Activity 3	Activity 4
In-house formal learning				
External courses				
Informal and personal				
Qualifications and/or experience gained?				

CHAPTER NINE

Designing your practice personal and professional development plan, using health and safety at work as an example[1]

The worked example demonstrates how you as a practice might set about preparing and undertaking your own programme. You might have chosen a more clinically based topic such as one or two commonly occurring occupational diseases or the whole field of occupational medicine in relation to the effects of work on your patients' health.

The stages in composing and undertaking your practice learning plan will include:

1 Identifying service development needs and staff learning needs using some of the range of methods in the first section of the previous chapter. Decide on the main areas of planned development for which you and other staff will need new knowledge and skills. Tie these in with the practice business plan (if you have not got one, you'd better start on that too!). Define:

- short-term objectives for the practice-based personal and professional development plan for the next year
- medium-term objectives for up to 3 years.

2 Identifying the learning needs that are essential to deliver your practice-based personal and professional development and clinical governance programmes – balance the clinical and non-clinical needs of individuals and their working environment (systems and procedures in the practice, the PCG/T, the NHS as a whole). This balance will include:

- generic learning that is relevant for everyone, e.g. communication
- team building
- specific skills for the particular roles and responsibilities of all included in the workplace-based plan.

You will already have undertaken much of the work of identifying your learning needs by carrying out the reflection exercises at the end of each chapter.

3 Assessing the infrastructure required to deliver your education and training plans for the practice and identifying from where you will obtain the necessary resources.

4 When making your overall practice-based learning plan you'll need to consider:

- the staff it involves: GPs, employed nurses and non-clinical staff; does it also include your cleaners, community pharmacists, attached staff and patients?
- the extent and resource costs of study time needed: actual costs to individuals and the practice, and also opportunity costs
- the practice's and practitioners' commitment
- the perspectives and needs of your staff, the practice as a whole, the PCG/T, the district and the wider NHS perspectives.

And how to:

- motivate the staff
- prioritise different learning needs between topics and between staff
- support staff and GPs
- evaluate what has been achieved
- assess and include new learning needs as they arise.

5 Devising the programme to meet the practice's prioritised learning needs; try to give each member of staff a definite role and responsibility to contribute to the overall practice-based plan. The programme should be written out as a timetabled action plan:

- a review of the success of previous year's plans and what is still outstanding
- how current learning needs will be identified
- who needs what and when; how needs will be met
- how team members will learn
- in-house teaching and learning

- reading
- meetings or courses outside the practice
- applying what has been learned by putting it into practice
- practice team work: awayday(s)
- how and by whom the learning will be evaluated and achievements monitored
- how and by whom new learning needs will be identified and included
- how any learning is to be disseminated.

6 Considering asking others from outside the practice to comment on whether they think you have framed your practice learning programme appropriately. You might ask patients, the public (i.e. non-users of your services), others in the PCG/T, local tutors, etc.

7 Making it happen – implement the practice personal and professional development action plan.

8 Evaluating the extent and quality of the service developments and associated learning; and the quality of the learning plan.

- Step 1: encourage individuals to self-appraise and evaluate their learning and contribute those reports to the overall evaluation.
- Step 2: evaluate the balance of the learning against the original objectives that were based on the practice business plan and service priorities.
- Step 3: use comments from patients and external individuals or bodies to evaluate the achievements and direction of the overall plan.
- Step 4: evaluate the practice-based learning plan itself – was it too ambitious, were there sufficient resources, were all the relevant staff covered, did all the GPs and staff join in?
- Step 5: how will you change or extend next year's plan?

9 Demonstrating that your working environment is fit for the GPs and practice staff to practise in with a clear record of what you have achieved: for example, improvements to the quality of patient care, staff well-being, effective systems, staff development.

10 Concluding what has still to be addressed and relay to drafting of the next practice PPDP.

11 Disseminating the achievements resulting from your practice-based PPDP.

The worked example on health and safety in the general practice is followed by a template for you to prepare your own plan – photocopy the pages – and get on with it! Demonstrate what you have achieved and keep a learning record.

Worked example of a practice PPDP: occupational healthcare of the primary healthcare team

What topic: health and safety at work in the practice
Who chose it? The practice manager suggested it. The rest of the practice readily agreed after the practice manager demonstrated how we were breaching the law in several instances.

Justify why topic is a priority:
a practice priority? Because at present there are breaches of health and safety law; and no proper records are kept.

a district priority? The PCG has designated 'risk assessment' and 'risk management' as a priority for practices – in respect of health and safety and other clinical topics. This will be an opportunity to develop expertise in risk management.

a national priority? Health and safety requirements are based on much national legislation.

Who will be included in our practice-based plan?
All GPs
Practice manager
Practice nurses
Reception staff
Cleaners
Attached staff who regularly use the practice premises – district nurses, community physio, health visitor, community midwives.

Who will collect the baseline information and how?
The practice manager will find out which regulations relating to health and safety apply to the practice; who can offer help and advice – for example any expert advisers at the health authority or local authority.

The practice manager will undertake the needs assessment – how the practice is faring compared with what is required (for example what policies are there, who knows about the policies, any risk assessments that have been done in the previous two years), identify gaps in compliance with the various regulations, extent of staff training, knowledge and skill base in the practice staff.

Where are we now?
The practice manager will undertake audits of what we are actually doing in our practice compared with the requirements of the main legislation as described in the example of a health and safety policy as described on page.... She will undertake a significant event audit of any recent breach in the law, or a patient or staff injury or accident – to see how systems and practices can be improved.

continued opposite

The practice manager will visit the general practice that the PCG nominates as a shining example to see how our practice compares and pick up ideas for making improvements; we will duplicate any of their policies and procedures that they are willing to share.

The practice manager will include awareness/knowledge/skills about health and safety matters in each member of staff's annual appraisal – making training a priority in each individual's action plan as appropriate.

What information will we obtain about individual learning wishes and needs?

Each member of staff will complete a checklist based on the self-appraisal checklist included in the section on learning needs assessment on page .... These will then be discussed with the practice manager as described above. The practice manager will put 'health and safety' on the agenda of the next staff meeting in the practice and invite comments and suggestions on improving health and safety at work – and volunteers to take responsibility for the various tasks.

What are the learning needs for our practice and how do they match the needs of the individual?

As the application of health and safety is a routine organisational issue rather than a clinical topic or interesting new subject that offers personal benefits, it is unlikely that individual members of staff will be clamouring to learn more about 'health and safety' for their own personal development. It is more likely that various staff will take on roles and responsibilities for the greater good of the practice in response to a direct request to do so.

Is there any patient or public input to our plan?[2]

We might use patient feedback to identify problems in the practice that need to be rectified to prevent mishaps. For instance a patient might tour the practice to point out 'accidents waiting to happen' such as a trailing wire, torn carpet, hazards within reach of children.

We might have an 'expert' in your patient population whom you could invite to help assess your situation and suggest improvements – for instance someone who specialises in ergonomically well-designed furniture.

How will we prioritise everyone's needs in a fair and open way?

In this instance the practice manager and GPs have to ensure that all the roles and responsibilities are allocated so that the regulations are fulfilled. It should be possible to decide for whom training is essential, for whom it is 'desirable' and who would like to attend out of interest rather than necessity. We might allocate the various duties at the practice meeting where health and safety is discussed; and the amount of education and training that is required by individual task-holders will follow.

continued overleaf

Objectives of practice PPDP arising from the preliminary data-gathering exercise

To increase the practice capability and expertise in health and safety matters such that we:

- comply with the regulations
- look after the well-being of the practice staff and promote fitness to work
- maintain a safe working environment for staff and patients
- minimise the effects of work on the health of the practice team members.

▼

Manual handling training session.

How we might integrate the 14 components of clinical governance[3] into our practice-based PDP focusing on the topic of health and safety at work in the practice

Establishing a learning culture: allocate roles and responsibilities for improving health and safety in line with legislation; identify everyone's learning needs and arrange training.

Managing resources and services: determine what changes need to be made to the practice environment to achieve minimum standards of health and safety; invest in necessary equipment, training, etc.

Establishing a research and development culture: undertake a survey to investigate problems in health and safety based on scientific principles.

continued opposite

Reliable and accurate data: monitoring health and safety and regular risk assessment requires good record keeping.

Evidence-based practice and policy: improving infection control will require searching for, and applying best practice in line with evidence-based health policies.

Confidentiality: tagging patients' records to indicate that they present a risk to the health and safety of staff (e.g. of violence) should be handled sensitively so that information is not released to anybody who does not have a 'need to know'.

Health gain: obvious health gains would result from for instance, reducing or avoiding risks for staff from a threat to their personal safety, or contamination from infection.

Coherent team: the successful application of a health and safety policy in general practice relies upon team members having a clear understanding of their roles and responsibilities.

Audit and evaluation: audit is an essential tool in monitoring standards of health and safety and in demonstrating effective risk management.

Meaningful involvement of patients and the public: a patient might walk around the surgery indicating any problems in health and safety, e.g. potential hazards.

Health promotion: good posture and hygienic practices are both example topics where health promotion would have potential benefits for staff.

Risk management: training in manual handling techniques for all those clinicians and non-clinical staff involved in lifting should reduce risks of injury at work from lifting. New staff will need to be trained as part of their induction.

Accountability and performance: revalidation will mean doctors not only demonstrating that they are fit to practise, but that their environment is fit to practise from. Demonstrating effective management of health and safety in the general practice workplace will be relevant here.

Core requirements: well-trained staff working in a healthy and safe environment should be cost-effective, as mistakes and accidents are expensive if patients and staff are affected.

Action plan
Agree who is involved/setting: as staff set out previously

Timetabled action: Start vv date

continued overleaf

By xx month: preliminary data gathering and collation of baseline of providers of advice/expertise, etc.:

- is there a practice protocol or guide on effective management of health and safety at work in the general practice; or other subsidiary protocols such as on stress management, personal safety, sickness absence?
- numbers of staff; map expertise; list other providers of advice/expertise outside the practice, e.g. from health authority, local NHS Trust; nominated leads on various aspects of health and safety
- information about past performance in practice – recent audits, or reports
- any relevant local and national priorities; and any additional associated resources for which we might apply, e.g. might be funding for upgrading premises
- staff discussion to report observed problems, views and suggestions for improvement.

by xy month: review current performance:

- practice manager reviews operation of services, e.g. is the safety policy for fire safety and fire prevention being applied?
- practice nurse checks that the first aid box is readily available and well stocked
- practice manager reviews extent of knowledge, skills and attitudes of practice team with respect to routine maintenance of health and safety in the practice
- audit actual performance versus pre-agreed criteria, e.g. look at accident book – have all incidents resulted in preventive action and been followed up to check that changes are in place and working?
- monitoring of staff and systems to check compliance with minimum standards/practice protocols
- compare performance with any or several of the 14 components of clinical governance, for example *risk management and learning culture would be very relevant.*

by yy month: identify solutions and associated training needs:

- set up new systems for reducing risks
- give practice team in-house training on important aspects of managing health and safety
- revise the practice protocols: address identified gaps in procedures; agree roles and responsibilities as a team for managing health and safety according to protocol; certain staff attend external courses.

by yz month: make changes:

- clinicians adhere to practice protocol, as shown by repeat audits; staff feedback
- staff trained at external courses share knowledge with others in practice at in-house training session with external facilitator as necessary
- organise further training to anticipate new requirements.

continued opposite

Expected outcomes: more effective management of health and safety; better staff compliance with practice protocols on health and safety; fewer or no accidents and injuries occurring in the general practice workplace; staff well-being promoted.

How does our practice PPDP tie in with other strategic plans?

The practice's business plan and the PCG/T's Primary Care Investment Plan might both prioritise achieving higher and more consistent standards in health and safety in the general practice workplace. The practice personal and professional development plan that focuses on health and safety would complement those strategic plans.

What additional resources will we require to execute our plan and from where do we hope to obtain them?

The practice might pay for the course fees of any member of staff undertaking training that fulfils a priority need of the practice, in this case, health and safety.

We may be able to justify an application for additional resources to our PCG/T or health authority or local NHS Trust using our preliminary learning and health needs assessments, tapping into the district or national strategic priorities.

If a member of staff is undertaking the training on behalf of the practice we should try to arrange that the training is undertaken in paid time. Any learning cascaded to other members of the practice team as part of the practice PPDP should also be undertaken in paid time and during working hours whenever possible.

How will we evaluate our practice PPDP?

We should be able to pick out methods of evaluation from the range of methods suggested for assessing learning needs in the first section of this chapter. The most appropriate methods will depend on what specific aims you set for your practice PPDP; for example if our main aim is that all staff lift safely, we might evaluate this by simply asking the community physiotherapist to observe staff at work. But if our aim was to improve the levels and appropriateness of education and information for staff about health and safety, we might evaluate our achievements by asking the staff themselves – by a simple test of knowledge, monitoring changes in practice, etc.

The practice manager and one of the GPs might plan the evaluation together and delegate the collection of data to a receptionist.

How will we know when we have achieved our objectives?

Usually this will be by comparing outcomes of your programme with baseline data. But it might also be determined by looking at staff compliance with the regulations as described in the practice protocol, or their levels of self-confidence in maintaining the aspects of health and safety for which they are responsible.

continued overleaf

How will we disseminate the learning from the plan to the rest of the practice team and patients? How will we sustain the new knowledge and skills?
We might write about it in a practice newsletter. Let all the staff know at practice meetings what progress has been made. We might want to describe our success at a PCG/T meeting or in a local report to the PCG/T.

We will review our practice protocol at set intervals to incorporate new information.

How will we handle new learning requirements as they crop up?
The practice manager might run audits at intervals and feed the results back to a practice meeting mid-way through the time period of the practice PPDP when there is time to revise the activities.

Record of our learning about health and safety at work

	Activity 1: application of health and safety policy	Activity 2: best practice in manual handling	Activity 3: improving personal safety	Activity 4: infection control
In-house formal learning	One-hour educational session facilitated by adviser on health and safety from local health promotion unit – all practice team involved to learn basic requirements; staff volunteered for various responsibilities	Community physiotherapist demonstrated best-practice in lifting techniques and good practice at same 1-hour educational session as for Activity 1		Practice nurse attends update on infection control – half-day; in treatment room and for individual consultations
External courses	Practice manager attended day course at regional venue – on legislation and effective management of health and safety in the NHS		One GP and two nurses attend local half-day course on improving personal safety	
Informal and personal	GPs chatted over coffee about implications of regulations whilst reading through the model health and safety policy the practice manager has drawn up	Practice manager pinned up on staff notice board pictures of how to lift/sit. Physio toured surgery with GP advising on hazards and need for new equipment to reduce risks from lifting and poor posture	GP/nurses review procedures to reduce risks to staff safety after course. Persuade other GPs to invest in better external lighting and alarms; new systems for keeping tabs on staff whereabouts	Practice nurses and practice manager discuss results of audit and need for new systems and procedures. Establish better recording systems; remove under blankets from couches, harbouring infection, etc.
Qualifications and/or experience gained?	Practice manager gained certificate of attendance at day course		PGEA accreditation for GP; certificate of attendance for nurses' portfolios	Certificate of course attendance; nurse inspects neighbouring practice and compares procedures to learn from each other

References

1   Wakley G, Chambers R and Field S (2000) *Continuing Professional Development in Primary Care: making it happen.* Radcliffe Medical Press, Oxford.

2   Chambers R (1999) *Involving Patients and the Public: how to do it better.* Radcliffe Medical Press, Oxford.

3   Chambers R and Wakley G (2000) *Making Clinical Governance Work for You.* Radcliffe Medical Press, Oxford.

Reflection and planning exercise

Photocopy the following template or complete the version in the book. Choose a topic that meets your individual needs, or duplicate the plan for the management of health and safety in the practice. You might choose to look at occupational healthcare of your patients as a wide-ranging topic; or you might focus down on one condition or an aspect such as risk management, or health and safety as given.

Justify the topic you have chosen for your practice team. Use some of the methods described to identify your learning needs. Incorporate learning needs or baseline information from 'Reflection exercises' at the end of each chapter. Draft the action plan. Discuss and agree it as a practice team as to whether it is relevant and well-balanced and achievable. Plan, obtain resources and undertake your learning activities. Demonstrate the subsequent improvements in the team's knowledge and practice.

It might take 10 to 30 hours to put the plan together and agree it as a practice team depending on what you do in the way of preliminary needs assessment and how many staff are involved.

Template for your practice PPDP: photocopy the pages and complete one chart per topic

What topic:

Who chose it?

Justify why topic is a priority:

A practice and professional priority?

A district priority?

A national priority?

Who will be included in the practice PPDP? (Give posts and names of GPs, employed staff, attached staff, others from outside the practice, patients?)

continued overleaf

Who will collect the baseline information and how?

Where are you now? (baseline)

What information will you obtain about individual learning wishes and needs? How will you obtain this and who will do it: self-completion check-lists, discussion, appraisal, audit, patient feedback?

What are the learning needs for the practice and how do they match the needs of the individual?

Is there any patient or public input to your plan?

How will you prioritise everyone's needs in a fair and open way?

continued opposite

Objectives of practice personal and professional development plan arising from the preliminary data-gathering exercise

How you might integrate the 14 components of clinical governance into your practice PPDP focusing on the topic of XXX?

Establishing a learning culture:

Managing resources and services:

Establishing a research and development culture:

Reliable and accurate data:

Evidence-based practice and policy:

Confidentiality:

Health gain:

Coherent team:

Audit and evaluation:

Meaningful involvement of patients and the public:

Health promotion:

Risk management:

continued overleaf

Accountability and performance:

Core requirements:

Action plan (include objectives, timetabled action, expected outcomes)

continued opposite

How does your practice-based learning plan tie in with your other strategic plans (for example the practice's business or development plan, the Primary Care Investment Plan)?

What additional resources will you require to execute your plan and from where do you hope to obtain them? (Will staff have to pay any course fees or undertake learning in their own time?)

How much protected time will you allocate to staff to undertake the learning described in your plan?

How will you evaluate your learning plan? (Who will be responsible for what?)

How will you know when you have achieved your objectives? (How will you measure success?)

continued overleaf

How will you disseminate the learning from your plan and sustain the developments and new-found knowledge or skills?

How will you handle new learning requirements as they crop up?

Record of your learning '   '; write in topic, date, time spent, type of learning

	Activity 1	Activity 2	Activity 3	Activity 4
In-house formal learning				
External courses				
Informal and personal				
Qualifications and/or experience gained?				

Sources of help

- BackCare. The National Organisation for Healthy Backs (registered as the National Back Pain Association), 16 Elmtree Road, Teddington, Middlesex TW11 8ST. Tel: 020 8977 5474. Fax: 020 8943 5318.
  email: 101540.1065@compuserve.com
  http://www.backpain.org

- Driver conditions
  http://www.dvlc.gov.uk

- DSS Benefits
  http://www.dss.gov.uk

- Employment Services
  http://www.dfee.gov.uk

- Faculty of Occupational Medicine, 6 St Andrew's Place, Regent's Park, London NW1 4LB. Tel: 020 7314 5890.
  http://www.facoccmed.ac.uk

- Health and Safety Executive Policy Unit, 8th Floor, South Wing, Rose Court, 2 Southwark Bridge, London SE1 9HS. Tel: 020 7717 6977. Fax: 020 7717 6417.

- Health and Safety Executive – main website
  http://www.hse.gov.uk

- HSE enquiries: contact HSE's Infoline Tel: 0541 545500 or write to HSE Information Centre, Broad Lane, Sheffield S3 7HQ.
  http://www.open.gov.uk/hse/hsehome.htm

- *Our Healthier Nation* website http://www.ohn.gov.uk

- Society of Occupational Medicine, 6 St Andrew's Place, Regent's Park, London NW1 4LB. Tel: 020 7486 2641.
  http://www.som.org.uk

Bibliography

Further reading, books, journal papers, relevant reports

Adisesh A and Parker G (1996) Working with an occupational health department. ABC of work related disorders. *BMJ.* **313**: 999–1002.

Baxter PJ (ed.) (1999) *Hunter's Diseases of Occupations* (9e). Arnold, London.

Chambers R (1997) *Occupational Health Services for GPs: a national model.* Royal College of General Practitioners/British Medical Association, London.

Chambers R and Campbell I (1996) Anxiety and depression in GPs: associations with type of practice, fundholding, gender and other personal characteristics. *Fam Pract.* **13**(2): 170–3.

Chambers R, Wall D and Campbell I (1996) Stresses, coping mechanisms and job satisfaction in general practitioner registrars. *Br J Gen Pract.* **46**: 343–8.

Chambers R, Miller D, Tweed P and Campbell I (1997) Exploring the need for an occupational health service for those working in primary care. *Occup Med.* **47**: 485–90.

Chambers R, George V, McNeill A and Campbell I (1998) Health at work in the general practice. *Br J Gen Pract.* **48**: 1501–4.

Committee on Health Promotion (1995) *Guidelines for Health Promotion Number 40.* Faculty of Public Health Medicine, Royal College of Physicians, London.

Cox RAF, Edwards FC and McCallum RI (eds) (2000) *Fitness for Work: the medical aspects* (3e). OUP, Oxford.

Department of Health (1994) *Occupational Health Services for NHS Staff.* HSG94 51. Department of Health, London.

Department of Health (1996) *Immunisation Against Infectious Disease.* HMSO, London.

Department of Health (1999) *The Healthy Workplace Initiative. Our Healthier Nation.* Department of Health, London.

Department of Social Security (2000) *A Guide for Registered Medical Practitioners.* Department of Social Security, London.

Department of Social Security (2000) *Medical Evidence for Statutory Sick Pay, Statutory Maternity Pay and Social Security Incapacity Benefit Purposes: a guide for registered medical practitioners.* Department of Social Security, London.

Farrow SC, Zeuner D and Hall C (1999) Improving infection control in general practice. *J Roy Soc Prom Health.* **119**(1): 17–22.

Health Education Authority (1999) *Framework for Action. Health at Work in the NHS.* Health Education Authority, London.

Health and Safety Commission (1997) *Violence and Aggression to Staff in the Health Services.* HSE Books, Sudbury.

Health and Safety Commission (1999) *Management of Health and Safety at Work. Approved Code of Practice.* L21. HSE Books, Sudbury.

Health and Safety Commission (2000) *Employee Consultation and Involvement in Health and Safety. Discussion document.* HSE Books, Sudbury.

Health and Safety Executive (1992) *A Guide to the Health and Safety at Work, etc. Act 1974* (5e). L1. HSE Books, Sudbury.

Health and Safety Executive (1993) *Your Patients and their Work. An introduction to occupational health for family doctors.* HSE Books, Sudbury.

Health and Safety Executive (1998) *Five Steps to Risk Assessment.* HSG 183. Health and Safety Executive, Sudbury.

Health and Safety Executive (1994) *Getting to Grips with Manual Handling.* INDG 143. HSE Books, Sudbury.

Health and Safety Executive (1996) *A Guide to the Health and Safety (Consultation with Employees) Regulations 1996.* L95. HSE Books, Sudbury.

Health and Safety Executive (1997) *Infections in the Workplace to New and Expectant Mothers.* HSE Books, Sudbury.

Health and Safety Executive (1998) *Safe Disposal of Clinical Waste.* HSE Books, Sudbury.

Health and Safety Executive (1998) *Developing an Occupational Health Strategy for Great Britain. Discussion document.* HSE Books, Sudbury.

Health and Safety Executive (1999) *An Introduction to Health and Safety: health and safety in small firms.* HSE Books, Sudbury.

Health and Safety Executive (1999) *Essentials of Health and Safety at Work.* HSE Books, Sudbury.

Jones R, Bly J and Richardson J (1990) A study of a work site health promotion program and absenteeism. *J Occ Med.* **32**(2): 95–9.

Litchfield P (ed.) (1995) *Health Risks to the Health Care Professional.* Royal College of Physicians, London.

Moore R and Moore S (1995) *Health and Safety at Work: guidance for general practitioners.* Royal College of General Practitioners, London.

Occupational Health Committee (1994) *The Occupational Physician.* British Medical Association, London.

Palmer K and Coggon D (1996) Investigating suspected occupational illness and evaluating the workplace. ABC of work related disorders. *BMJ.* **313**: 809–11.

Parker G (1996) General practitioners and occupational health services. *Br J Gen Pract.* **46**: 303–5.

Royal College of Nursing (1995) *Health Assessment: advice for occupational health nurses.* Royal College of Nursing, London.

Royal College of Nursing (1995) *Health Assessment: advice to managers.* Royal College of Nursing, London.

Seaton A, Agius R, McCloy E and D'Auria D (1994) *Practical Occupational Medicine.* Arnold, London.

Sen D and Osborne K (1997) General practices and health and safety at work. *Br J Gen Pract.* **47:** 103–4.

Stacey NH (ed.) (1993) *Occupational Toxicology.* Taylor and Francis, London.

Stationery Office (1998) *Our Healthier Nation: a contract for health. CM3852.* The Stationery Office, London.

Stationery Office (1998) *Working Together for a Healthier Scotland. CM3854.* The Stationery Office, London.

Stationery Office (1998) *Better Health – Better Wales. CM 3922.* The Stationery Office, London.

von Onciul J (1996) Stress at work. ABC of work related disorders. *BMJ.* **313**: 745–8.

HSE leaflets (*see* Catalogue 34 published in 1999 – all HSE Books, Sudbury unless otherwise specified)

HSE leaflets, pamphlets and books may be obtained from booksellers or by mail order from: HSE Books, PO Box 1999, Sudbury, Suffolk CO10 6FS. Tel: 01787 881165; Fax: 01787 313995.

HSE (1989) *Health & Safety Law: what you should know* [leaflet pack of 50].

HSE (1989) *Safety of Pressure Systems: pressure systems and transportable gas containers regulations.* COP37.

HSE (1990) *First Aid at Work: Health & Safety (First Aid) Regulations 1981* (Approved Code of Practice). L74.

HSE (1991) *Seating at Work.* HSG57.

HSE (1992) *Management of Health & Safety at Work* (Approved Code of Practice). L21.

HSE (1998) *Safe Use of Work Equipment (Approved Code of Practice and Guidance).* L22.

HSE (1998) *Manual Handling Guidance on Regulations.* L23.

HSE (1992) *Workplace Health, Safety & Welfare Approved Code of Practice.* L24.

HSE (1992) *Personal Protective Equipment at Work Guidance on Regulations.* L25.

HSE (1992) *Display Screen Equipment Work Guidance on Regulations.* L26.

HSE (1993) *Your Work and Your Health: what your doctor needs to know.* INDG116.

HSE (1993) *Writing Your Health & Safety Policy Statement: a guide to preparing a safety policy statement for a small business.*

HSE (1993) *Electricity at Work: safe working practices.* HSG85.

HSE (1993) *A Step by Step Guide to COSHH Assessment.* HSG97.

HSE (1994) *VDUs: an easy guide to the Regulations.* HSG90.

HSE (1994) *Upper Limb Disorders: assessing the risks.* INDG171.

HSE (1994) *You can do it.* [poster]

HSE (1995) *Stress at Work: a guide for employers.* HSG116.

HSE (1995) *Health Risk Management: a practical guide for managers in small and medium sized enterprises.* HSG137.

HSE (1995) *Respiratory Sensitisers and COSHH.* INDG95.

HSE (1995) *Passive Smoking at Work.* INDG63.

HSE (1995) *Workplace Health, Safety and Welfare: a short guide.* INDG170.

HSE (1995) *Read the Label: how to find out if chemicals are dangerous.* INDG186.

HSE (1996) *Slips and Trips: guidance for employers on identifying hazards and controlling risks.* HSG155.

HSE (1997) *Successful Health & Safety Management.* HSG65.
The Disability Discrimination Act (1995). The Stationery Office, London.

HSE (1999) *A Guide to the Reporting of Injuries, Diseases and Dangerous Occurrences Regulations 1995.* (Revised) L73.

HSE (1996) *Violence at Work.* INDG69.

HSE (1999) *'Good Health is Good Business': employer's guide.* MISC130.

HSE (1999) *COSHH: a brief guide to the Regulations.* (Revised) INDG136. (Pack of 10 leaflets.)

Appendix 1
Health and safety legislation: extracts of key requirements of legislation

You may like to obtain copies of key legislation for reference in your own workplace. We would suggest that you obtain copies of:

Health and Safety at Work, etc. Act 1974
Management of Health and Safety at Work Regulations and ACOP* 1999
Health and Safety (Display Screen Equipment) Regulations and ACOP 1992
Control of Substances Hazardous to Health Regulations and ACOPs 1999
Health and Safety (First Aid) Regulations 1981 and Guidance.

(*Approved Code of Practice)

These all contain important information on *managing* health and safety at work, and will give you an idea of the way health and safety legislation is moving.

Health and Safety at Work, etc. Act 1974 – extracts

Health and Safety at Work, etc. Act 1974. HMSO, London.

> s.2 'It shall be the duty of every employer to ensure, so far as is reasonably practicable, the health, safety and welfare at work of all his employees ...'

> s.3 'It shall be the duty of every employer to conduct his undertaking in such a way as to ensure, so far as is reasonably practicable, that persons not in his employment who may be affected thereby are not thereby exposed to risks to their health or safety ...'

> s.7 'It shall be the duty of every employee while at work ... to take reasonable care for the health and safety of himself and of other persons who may be affected by his acts or omissions at work ...'

Management of Health and Safety at Work Regulations and Approved Code of Practice 1999 – extracts

Regulation 5 Health Surveillance:

> Every employer shall ensure that his employees are provided with such health surveillance as is appropriate having regard to the risks to their health and safety which are identified by the assessment.

Health and Safety (Display Screen Equipment) Regulations and Approved Code of Practice 1992 – extracts

Regulation 5 Eye and eyesight:

> Where a person [is a user of display screen equipment] his employer shall ensure that he is provided with an appropriate eye and eyesight test, any such test to be carried out by a competent person ...

> Display screen equipment users are not obliged to have such tests performed, but where they choose to exercise their entitlement, employers should offer an examination by a registered ophthalmic optician, or a registered medical practitioner with suitable qualifications ... (All registered medical practitioners, including those in company occupational health departments, are entitled to carry out sight tests but normally only those with an ophthalmic qualification do so.)

The Reporting of Injuries, Diseases and Dangerous Occurrences Regulations 1995 (RIDDOR) – extracts

RIDDOR requires the reporting of work-related accidents, diseases and dangerous occurrences. It applies to all work activities, but not to all incidents. You (the employer) are required to report to the HSE or Local Authority if there is an accident at work and:

- your employee or a self-employed person working on your premises is killed, or a member of the public is killed or taken to hospital, or

- one of your employees or a self-employed person working on your premises suffers an injury from an accident or an act of violence resulting in them being absent from (or unable to do) their normal work for more than three days including non-working days, or
- a doctor notifies you that your employee is suffering from a reportable work-related disease, or
- if something happens which does not result in a reportable injury, but which clearly could have done (a 'dangerous occurrence').

The terms 'major injury' and dangerous occurrence are defined in the guide to the Regulations.

Appendix 2
Occupational Health Advisory Committee (2000)
Improving Access to Occupational Health Support: extracts

Occupational Health Advisory Committee (2000) *Improving Access to Occupational Health Support.* HSE Books, Sudbury. Downloadable from the HSE's website. Reproduced here by kind permission of the Health and Safety Executive.

Conclusions (s1.5)

The main overarching conclusions reached by the Occupational Health Services Working Group (OHSWG) which have informed all its recommendations are:

(a) prevention of ill health at work and amelioration of the effects of health on work, e.g. through rehabilitation, are essentially management issues and whilst professional occupational health support may be required, this is not inevitably the case in all circumstances. The key will be to ensure that employers and managers have access to a point of enquiry that can either suggest solutions or signpost employers and managers to the appropriate level of advice

(b) strategies to remove occupational health inequalities and improve access to occupational health support will not succeed, unless further action is taken to improve employer and worker awareness of when such support is needed

(c)   delivery mechanisms for occupational health support should give priority to the prevention of health risks at work and the issues that arise from the effects of health on work, e.g. non-work-related illness compounded by work, and rehabilitation

(d)   there is no one solution that will meet the occupational health support needs of everyone; flexibility is the key to delivery mechanisms

(e)   there is a wide range of mechanisms, many involving partnerships, that should be pursued to raise awareness of occupational health issues, and encourage and facilitate the delivery and use of occupational health support.

The meaning of 'occupational health' and 'occupational health support' (s3.1–3.4)

s3.2. The term 'occupational health' conveys different things to different people. For some it means simply the prevention and treatment of illness that is directly related to work, in which health education has no place. Others will emphasise fitness for work issues separately from health and safety. However the impact of any sickness absence on small and medium enterprises (SMEs) and their employees does not brook such fine distinctions. In those terms, it is of little importance whether the sickness is the result of an accident at work, long-term exposure to risk or the conflicting demands of home and work. Only a holistic approach can make a difference to health inequalities. The recommendations of this report are therefore based on the broad view that occupational health can embrace:

(f)   the effect of work on health, whether through sudden injury or through long-term exposure to agents with latent effects on health, and the prevention of occupational disease through techniques which include health surveillance, ergonomics, and effective human management systems

(g)   the effect of health on work, bearing in mind that good occupational health practice should address the fitness of the task for the worker, not the fitness of the worker for the task alone

(h)   rehabilitation and recovery programmes

(i)   helping the disabled to secure and retain work

(j)   managing work-related aspects of illness with potentially multifactorial causes (e.g. musculoskeletal disorders, coronary heart disease) and helping workers to make informed choices regarding lifestyle issues.

s3.3. The OHSWG included the term 'occupational health support' in its title but it soon became clear that this was too narrow a description of the kind of help that employers needed to fulfil in order to protect the health of their employees. Although there is no generally accepted definition of what constitutes an occupational health service and no clear single form of practice, such services are traditionally understood to be medically based and led by doctors or nurses. Many existing providers of occupational health services, particularly in the private sector are now able to field multidisciplinary teams including, for example, occupational hygienists, ergonomists, psychologists, health and safety specialists and counsellors, etc. But such teams are still likely to be led by a doctor or nurse. However, what many SMEs need in practice is simple, sector-specific guidance on practical measures to reduce exposure to hazardous agents, advice on enabling workers with health problems to continue working, together with information they can pass to their employees about ways of keeping healthy. Such advice may be obtainable from engineers and other technicians, trade associations, suppliers of material and equipment and safety representatives. Many workers in small businesses are likely to remain reliant on primary care, trade unions, and safety representatives for advice. Therefore, this report uses the expression 'occupational health support' throughout to indicate the full range of advice that SMEs and workers may need to tap into.

s3.4. Good occupational health practice (as fostered by good occupational health support) is one of the factors that should lead, in the longer term, to positive outcomes for workers and businesses alike, in terms of a good quality of life inside and outside work, the social and material advantages of work, reduced sickness absence, higher productivity, a good, responsible image for individual businesses and greater national wealth creation.

Current legislation

s5.2. The Health and Safety at Work, etc. Act (HSAWA) places a wide-ranging duty on employers to protect the safety, health and welfare of their employees. Regulations made under the HSAWA and other legislation place specific duties on employers relating to, for example, risk assessment, health surveillance, managing health in certain sectors, fitness for work in occupations entrusted with public safety, protecting the vulnerable and employing the disabled. However ... none of these provisions place a duty on employers to buy in or provide occupational health services. Instead the Management of Health and Safety at Work Regulations 1992 require employers (with certain exceptions) to appoint competent persons to fulfil their statutory responsibilities ... the ... Regulations ... make it clear that the preferred way of complying is to appoint people from within the workforce ...

s5.3. Consultations on occupational health problems need not always be external, or conducted with specialists ... The Safety Representatives and

Safety Committees Regulations 1997 and the Health and Safety (Consultation with Employees) Regulations 1996 require employers to consult with employees ...

s5.4. HSAWA also places duties on employees to take reasonable care for the health and safety of themselves and others ...

s5.5. The Disability Discrimination Act 1995 (DDA) requires employers with 15 or more employees to treat disabled persons in all employment matters and make any reasonable changes to the premises, job design, etc., that may be needed to accommodate the needs of disabled. DDA extends the definition beyond deafness, blindness, mental illness and physical impairment, to include severe disfigurement as well as progressive conditions such as AIDS where disability develops some time after first diagnosis ...

Priority needs of small and medium enterprises (SMEs)

s8.8. There is a general consensus ... that the following needs should be addressed:

(a)　the overriding need to educate employers and raise their awareness of:
- the existence of health problems for which competent advice may be needed
- how to obtain that advice
- how to ensure that competent persons appointed from within the workforce are working within their knowledge and experience and when to seek further help
- the need for an integrated approach to risk management
- ways of using occupational health support and advice to create a healthy working environment
- ways of complying with legislation
- ways of involving others in the control of risks

(b)　help with hazard identification, risk assessment and implementation of controls ...

(c)　practical help with prevention of ill-health through the use of tools such as monitoring of sickness absence, health surveillance, and systems that allow a balanced flow of work

(d)　help with human resource issues, good management skills and flexible working policies, and healthy living issues

(e)   advice on fitness for work issues, redeployment, rehabilitation and employing the disabled
(f)   help with setting up first aid, and where appropriate other treatment services ...
(g)   provision of a local one-stop shop approach to business advice including occupational health and safety.

The needs of larger companies

s9.1. Although employees of larger companies are more likely to have access to occupational health advice, lack of awareness of occupational health and suitable training remains a problem in some companies and economic pressures may lead to a reduction in, e.g. health surveillance and the monitoring of control measures.

The needs of workers and others

s10.2. Workers should be encouraged to take more responsibility for their own health at work. To do this they need to have confidence and understanding that:

(a)   occupational ill-health can be prevented and solutions to occupational health problems are available
(b)   people with disabilities and health problems can work provided that adjustments are made and work is designed to suit the individual
(c)   with advice and training, workers and their representatives can contribute significantly to the control and prevention of risks to health.

s10.3. Priority needs for workers include:

(a)   occupational health support that is seen to be objective and independent of undue employer/management influence, ethical and of the highest probity
(b)   support and advice concerned with risk prevention, disease prognosis and referral for treatment in a way that is not prejudicial to income or job security
(c)   where appropriate, support that is tailored to the needs of peripatetic workers and people who work at or from home ...
(d)   training to allow safety representatives and other workers to become involved in risk prevention strategies and
(e)   the means of ensuring that their occupational health history is not 'lost' after job moves.

Synopsis of recommendations: seven of the 30 recommendations are given here

Recommendation 1: Comprehensive frameworks for occupational health support should be established that will raise awareness and enable access to such support for everyone who needs it.

Recommendation 2: Occupational health resources and delivery mechanisms should give priority to the prevention of health risks at work and the issues that arise from the effects of health on work, e.g. non-work-related illness compounded by work, and rehabilitation.

Recommendation 3: It should be mandatory for occupational health to form one of the core issues to be covered by HImPs.

Recommendation 19: GPs and practice nurses should be encouraged to take up formal and informal opportunities for training and experience in occupational health, including distance learning, membership of learned societies, opportunities for site visits, and networking with local NHS occupational health units. The feasibility of training other primary care staff to provide occupational health advice should be explored.

Recommendation 20: Primary Care Groups (or Trusts) should be strongly encouraged to take on board occupational health needs and engage someone with occupational health and safety expertise within the PCG/T.

Recommendation 22: The DoH/NHS Executive should examine the feasibility of developing and extending the provision of occupational health units within the NHS without prejudicing the occupational health support provided for NHS staff.

Recommendation 23: Consideration should be given to the benefits of recognising occupational health as an NHS specialty to which primary care patients can be referred.

Index